Planar Graphs

Titles in This Series

9 **William T. Trotter, Editor,** Planar Graphs

8 **Simon Gindikin, Editor,** Mathematical Methods of Analysis of Biopolymer Sequences

7 **Lyle A. McGeoch and Daniel D. Sleator, Editors,** On-Line Algorithms

6 **Jacob E. Goodman, Richard Pollack, and William Steiger, Editors,** Discrete and Computational Geometry: Papers from the DIMACS Special Year

5 **Frank Hwang, Clyde Monma, and Fred Roberts, Editors,** Reliability of Computer and Communication Networks

4 **Peter Gritzmann and Bernd Sturmfels, Editors,** Applied Geometry and Discrete Mathematics, The Victor Klee Festschrift

3 **E. M. Clarke and R. P. Kurshan, Editors,** Computer-Aided Verification '90

2 **Joan Feigenbaum and Michael Merritt, Editors,** Distributed Computing and Cryptography

1 **William Cook and Paul D. Seymour, Editors,** Polyhedral Combinatorics

DIMACS

Series in Discrete Mathematics and Theoretical Computer Science

Volume 9

Planar Graphs

William T. Trotter
Editor

NSF Science and Technology Center
in Discrete Mathematics and Theoretical Computer Science
A consortium of Rutgers University, Princeton University,
AT&T Bell Labs, Bellcore

American Mathematical Society

This DIMACS volume resulting from the Special Year on Graph Theory and Algorithms contains research articles and extended abstracts from participants at the Planar Graphs Workshop held at DIMACS from November 18, 1991 through November 21, 1991.

1991 *Mathematics Subject Classification*. Primary 05C10, 05C35, 05C38, 05C75, 05C85.

Library of Congress Cataloging-in-Publication Data

Planar graphs/William T. Trotter, editor.
p. cm.—(DIMACS series in discrete mathematics and theoretical computer science; v. 9)
"Research articles and extended abstracts of participants at the Planar Graphs Workshop held at DIMACS from November 18, 1991 through November 21, 1991"—Fwd.
Includes bibliographical references.
ISBN 0-8218-6600-1
1. Graph theory—Congresses. I. Trotter, William T. II. Series.
QA166.P56 1993 93-22339
511′.5–dc20 CIP

Printed in the United States of America.
The paper used in this book is acid-free and falls within the guidelines established to ensure permanence and durability. ♾

This volume was prepared by the authors using AMS-TEX and AMS-LATEX, the American Mathematical Society's TEX macro system.

10 9 8 7 6 5 4 3 2 1 98 97 96 95 94 93

Contents

Foreword

This DIMACS volume resulting from the Special Year on Graph Theory and Algorithms contains research articles and extended abstracts of participants at the Planar Graphs Workshop held at DIMACS from November 18, 1991 through November 21, 1991. We especially thank the Workshop organizer, Tom Trotter, for organizing an activity that brought together so many outstanding speakers and participants dealing with this important field.

Fred Roberts, Acting Director
Robert Tarjan, Co-Director
Diane Souvaine, Associate Director

Preface

During the 1991/92 academic year, Fan R. K. Chung and I served as Co-Directors of the Special Year on Graph Theory and Algorithms sponsored by the Center for Discrete Mathematics and Theoretical Computer Science (DIMACS). Based on informal feedback from some of the more than 300 researchers who traveled to New Jersey this past year to participate in DIMACS activities, the Special Year was a great success. This admittedly biased view is supported by the growing list of accomplishments in which researchers recognize the role played by DIMACS in initiating, fostering and facilitating their work.

The Special Year on Graph Theory and Algorithms featured four workshops:

- Random Graphs and Randomized Algorithms
  Milena Mihail
  Joel Spencer
- Planar Graphs: Structures and Algorithms
  Tom Trotter
- Graph Embeddings and Parallel Architectures
  Tom Leighton
  Bruce Maggs
  Arny Rosenberg
- Expander graphs: Theory and Applications
  Persi Diaconis
  Joel Friedman

This monograph consists of research articles and extended abstracts submitted by participants in the Workshop on Planar Graphs. More than 70 persons attended this Workshop, and these papers cover a wide range of topics including: enumeration, characterization problems, algorithms, extremal problems, network flows and geometry.

I would like to express my appreciation to the persons who attended the Planar Graphs Workshop and contributed to its success. Also, special thanks to the authors of the articles appearing in this volume and to the many anonymous referees for their assistance with the composition of a volume which will (hopefully) have lasting scientific value. Finally, thanks to Christine Thivierge, Donna Harmon and Nancy Gerstl for their editorial and secretarial assistance.

Tom Trotter

Director, Combinatorics and Optimization
Bell Communications Research

Regents Professor
Arizona State University

DIMACS Series in Discrete Mathematics
and Theoretical Computer Science
Volume 9, 1993

Cycles, Cocycles and Diagonals: A Characterization of Planar Graphs

DAN ARCHDEACON, C. P. BONNINGTON, AND C. H. C. LITTLE

September 23, 1992

ABSTRACT. We give a characterization of planar graphs in terms of a set of walks which contain each edge exactly twice and which interacts with the cycle and cocycle spaces of the graph in a special manner.

Extended Abstract

Let G be a graph. A *polygon* in G is the edge set of a connected 2-regular subgraph. The polygons of G span a vector space over $\mathbb{Z}_2$, called the *cycle space,* in which addition of sets is defined as their symmetric difference. The elements of the cycle space are called *cycles.* Thus a cycle is the edge set of a subgraph of G in which the degree of every vertex is even. The orthogonal complement of the cycle space (that is, the collection of edge sets whose intersection with every cycle is of even cardinality) is called the *cocycle space,* and its elements are *cocycles.* Thus a set S of edges is a cocycle if and only if there is a set T of vertices such that S is the set of edges joining a vertex of T to a vertex of $V(G) - T$. A set of edges which is both a cycle and a cocycle is called a *bicycle.* An example is the empty set.

Rosenstiehl and Read [**RR**] have shown the following striking theorem which classifies the edges of an arbitrary graph into three types.

THEOREM 1. *For any edge e in a graph G, exactly one of the following holds:*

(1) *e belongs to a cycle C for which $C - \{e\}$ is a cocycle,*
(2) *e belongs to a cocycle C for which $C - \{e\}$ is a cycle, or*
(3) *e belongs to a bicycle.*

1991 *Mathematics Subject Classification.* Primary 05C10, 05C75.

The first author is partially supported by NSF grant number DMS-9007503.
This is an extended abstract of a paper which has been submitted for publication elsewhere.

We call an edge *cyclic, cocyclic*, or *bicyclic* according to whether it satisfies Conditions (1), (2), or (3) above respectively.

Suppose that C is a cycle containing an edge e with $C - \{e\}$ a cocycle. Then we call C a *principal cycle* for e. A *principal cocycle* for e is defined similarly. Principal cycles and principal cocycles for a given edge are not unique, unless the only bicycle in G is empty.

A *walk* W is an ordered set $v_0e_1v_1e_2v_2\dots e_nv_n$ where each v_i is a vertex and each e_i is an edge joining v_{i-1} to v_i. The walk is *closed* when $v_0 = v_n$. For any walk W we define the *residue* $\nabla(W)$ of W as the set of edges which appear an odd number of times.

Now suppose that G is a connected graph with no non-empty bicycles, and let D be a walk in which each edge appears exactly twice. For a given edge e we distinguish two cases. Suppose first that D is of the form $uevAuevBu$ where A and B are walks. Then vAu and vBu are walks called the *halves* of D determined by e. Their residues are equal, since $\nabla(D) = \emptyset$. We denote this residue by D_e. In the remaining case D is of the form $uevAveuBu$ where A and B are walks. This time the halves of D determined by e are defined as $uevAv$ and $veuBu$ (the idea being that a half should join the ends of e). Again their residues are equal and denoted by D_e.

We call D an *algebraic diagonal* of G if, for each edge e, D_e is a principal cocycle for e. Thus $D_e + \{e\}$ is the principal cycle for e. This concept was used by Rosenstiehl and Read [**RR**] to formulate the following criterion for the planarity of a graph with no non-empty bicycles.

THEOREM 2. *Let G be a connected graph with no non-empty bicycles. Then G has an algebraic diagonal if and only if G is planar.*

We now ask "What is the proper generalization of an algebraic diagonal which gives the analogous theorem characterizing planar graphs with bicycles?" It turns out that our diagonals may contain more than one walk. Accordingly we need a more general concept than the halves of a single walk determined by an edge in order to keep track of the principal cycles and cocycles. When talking about a family of walks we do not wish to distinguish a starting point or a direction on a closed walk. We define a *tour* as an unrooted undirected closed walk.

Let G be a graph and let $\mathcal{W}$ be a family of tours which collectively use each edge of G twice. Such a family will be called a *double cover.* A *ladder* $\mathcal{L}$ in $\mathcal{W}$ is a cyclic sequence $e_1W_1e_2W_2\dots e_lW_l$ where each W_i is a tour in $\mathcal{W}$ and each e_i is a link (an edge which is not a loop) in both tours W_{i-1} and W_i and distinct from each other e_j. A ladder is *simple* if no two tours are equal. As an example, a simple ladder in K_4 has tours given by the three Hamiltonian polygons and links the three edges incident on a given vertex.

Fix a preferred direction on each tour W_i in a ladder $\mathcal{L}$. The edge e_i in W_{i-1} and W_i is *consistent* if traversed twice in one direction in both those tours, and *inconsistent* otherwise. If the tours can be directed so that each e_i is inconsistent, then the ladder is *straight;* otherwise the ladder is *twisted.* The example above is twisted.

Our definition of a diagonal will involve properties not just of a single tour, but of ladders. In particular we need some distinguished sets of edges in a ladder to form a cocycle. To this end direct each W_i and write it as $e_iP_ie_{i+1}P'_i$ where P_i and P'_i are walks. A *side* of the ladder is the collection $\cup_{i=1}^{l} P_i$ together with those e_i's which are inconsistent for the given directions. Thus a side of a ladder is not uniquely defined, but depends on the directions assigned to the tours.

A *diagonal* $\mathcal{D}$ in a graph G is a double cover such that the residue of each tour in $\mathcal{D}$ and the residue of any side of each simple ladder in $\mathcal{D}$ are cocycles. It can be shown that in a diagonal the residue of one side of a ladder $\mathcal{L}$ is a cocycle if and only if the residue of every other side of $\mathcal{L}$ is a cocycle. For example, the three Hamiltonian polygons form a diagonal of K_4. We can now formulate the following theorem, whose proof will appear elsewhere [**ABL**].

THEOREM 3. *A graph is planar if and only if it has a diagonal.*

The idea of the proof of sufficiency is to use Wagner's characterization of planar graphs. Specifically, we show that K_5 and $K_{3,3}$ do not have diagonals, and that if a graph G has a diagonal then so do its minors. Necessity is shown by establishing that the left-right walks of a planar embedding of a graph constitute a diagonal.

It can be easily shown that Theorem 2 follows from Theorem 3.

ACKNOWLEDGMENT. The third author gratefully acknowledges the support of the University of Vermont, where he was a Visiting Professor while much of this research was done.

REFERENCES

[ABL] Dan Archdeacon, C.P. Bonnington, and C.H.C. Little, *An Algebraic Characterization of Planar Graphs*, submitted.

[RR] P. Rosenstiehl and R. Read, *On the Principal Edge Tripartition of a Graph*, Annals of Discrete Math. **3** (1978), 195–226.

DEPARTMENT OF MATHEMATICS AND STATISTICS, UNIVERSITY OF VERMONT, BURLINGTON, VT, USA 05405
E-mail address: archdeac@uvm.edu

DEPARTMENT OF MATHEMATICS, UNIVERSITY OF WAIKATO, HAMILTON, NEW ZEALAND
E-mail address: bonning@mat.aukuni.ac.nz

DEPARTMENT OF MATHEMATICS, MASSEY UNIVERSITY, PALMERSTON NORTH, NEW ZEALAND
E-mail address: C.Little@massey.ac.nz

DIMACS Series in Discrete Mathematics
and Theoretical Computer Science
Volume 9, 1993

Stack and Queue Layouts of Directed Acyclic Graphs

LENWOOD S. HEATH, SRIRAM V. PEMMARAJU AND ANN TRENK

December 9, 1992

ABSTRACT. Stack layouts and queue layouts of *undirected* graphs model problems in fault tolerant computing, in VLSI design, and in managing the flow of data in a parallel processing system. In certain applications, such as managing the flow of data in a parallel processing system, it is more realistic to use layouts of *directed acyclic graphs* (dags) as a model. A *stack layout* of a dag consists of a topological ordering σ of the nodes of the graph along with an assignment of arcs to stacks such that if the nodes are laid out in a line according to σ and the arcs are all drawn above the line, then no two arcs that are assigned to the same stack cross. A *queue layout* is defined analogously, except that arcs are assigned to queues with the condition that no two arcs assigned to a queue nest. The *stacknumber* of a dag is the smallest number of stacks required for its stack layout, while the *queuenumber* of a dag is the smallest number of queues required for its queue layout. Classes of dags identified by the structure of their underlying undirected graphs are studied. We determine the stacknumber and queuenumber of classes of dags whose underlying undirected graphs are trees or unicyclic graphs. We give a forbidden subgraph characterization of those dags with queuenumber equal to 1 whose underlying undirected graphs are trees. We discuss classes of planar dags and outerplanar dags that have queuenumber and stacknumber that differ markedly. Finally, we show that the problems of determining whether a dag can be laid out in 7 queues and of determining whether a dag can be laid out in 9 stacks are both NP-complete.

1991 *Mathematics Subject Classification.* 68R10; Secondary 05C10.

The research of the first and second authors was partially supported by National Science Foundation Grant CCR-9009953.

The third author was supported by an Eliezer Naddor Postdoctoral Fellowship in Mathematical Sciences from The Johns Hopkins University during the year 1991-1992 while in residence at Dartmouth College.

This is an extended abstract of a paper which will be submitted for publication elsewhere.

1. Definitions and Notation

A *k-stack layout* of a directed acyclic graph (a dag) $\vec{G} = (V, \vec{E})$ consists of a topological ordering σ of the nodes of $\vec{G}$ and an assignment of each arc of $\vec{G}$ to one of k stacks $s_1, s_2, \dots, s_k$. Each stack s_j obeys the last-in/first-out discipline and operates as follows. The nodes in V are scanned in the order given by σ. When a node v is encountered, any arcs assigned to s_j that have node v as their head must be at the top of the stack and are popped. Any arcs that are assigned to s_j and have node v as their tail, are pushed onto s_j. Arc (v, w) is pushed onto s_j *before* arc (v, x) if w occurs *after* x in the ordering σ. The *stacknumber* $SN(\vec{G})$ of $\vec{G}$ is the smallest k such that $\vec{G}$ has a *k-stack layout.*

A *k-queue layout* of a dag $\vec{G} = (V, \vec{E})$ consists of a topological ordering σ of the nodes of $\vec{G}$ and an assignment of each arc of $\vec{G}$ to one of k queues $q_1, q_2, \dots, q_k$. Each queue q_j obeys the first-in/first-out discipline and operates as follows. The nodes in V are scanned in the order given by σ. When a node v is encountered, any arcs assigned to q_j that have node v as their head must be at the front of the queue and are dequeued. Any arcs that are assigned to q_j and have node v as their tail, are enqueued onto q_j. Arc (v, w) is enqueued on q_j *before* arc (v, x) if w occurs *before* x in the ordering σ. The *queuenumber* $QN(\vec{G})$ of $\vec{G}$ is the smallest k such that $\vec{G}$ has a *k-queue layout.*

The *stacknumber* $SN(\mathcal{C})$ of an infinite class of dags $\mathcal{C}$ is the function of the integer n that equals the least upper bound on the number of queues necessary to lay out all graphs in $\mathcal{C}$ having at most n vertices. The *queuenumber* $QN(\mathcal{C})$ of $\mathcal{C}$ is defined analogously.

2. Background and Motivation

Stack layouts and queue layouts of undirected graphs have appeared in a variety of contexts such as VLSI, fault-tolerant computing, parallel processing, and sorting networks. Bernhart and Kainen [**1**] introduce the concept of a stack layout, calling it a *book embedding.* Motivated by problems in VLSI and fault-tolerant processing, Chung, Leighton, and Rosenberg [**2**] study stack layouts of undirected graphs and provide optimal stack layouts for a variety of classes of graphs. Motivated by problems in scheduling processes in a parallel processing system, Heath and Rosenberg [**6**] develop the notion of queue layouts and provide optimal queue layouts for many classes of undirected graphs. Heath and Pemmaraju [**4**] use queue layouts of graphs as a basis for a scheme that efficiently performs matrix computations on a systolic array.

In some applications of stack and queue layouts, it is more realistic to model the application domain with directed acyclic graphs (dags), rather than with undirected graphs. One such application is the problem of process scheduling on a system of parallel processors [**8**]. A process can be viewed as a set of computations that have certain interdependencies. Therefore, a process can be modeled by a dag, called the *dependency dag*, with nodes representing computations and

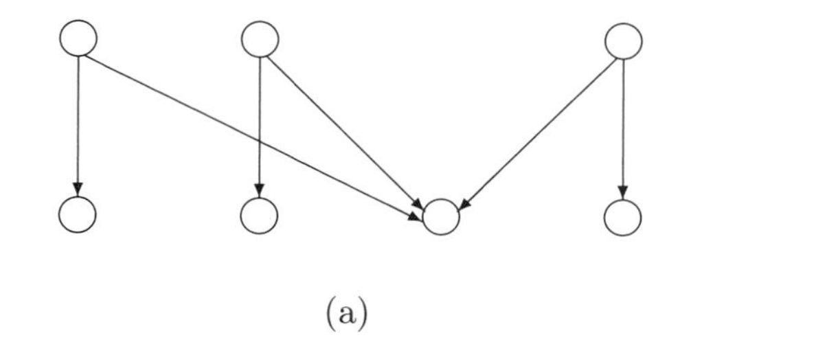

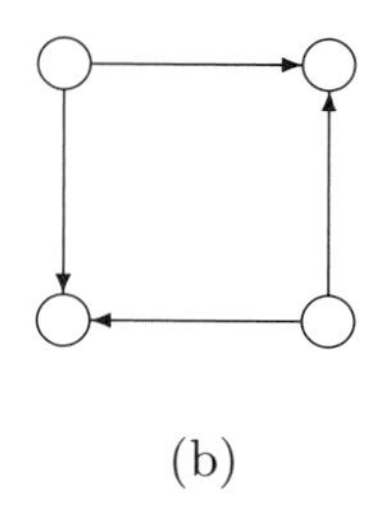

FIGURE 1. (a) A directed tree with queuenumber 2. (b) A unicyclic dag with queuenumber and stacknumber 2.

arcs representing dependencies. In a parallel processor system, computations are assigned to different processors so that independent computations can be performed in parallel, with the goal of reducing the total computation time. The processors to which computations have been assigned need to exchange data as dictated by the interdependencies among the computations. This flow of data can be managed by a set of queues (or stacks) as explained below. Each arc (dependency) is assigned to a queue, with the following meaning. If arc (u, v) is assigned to queue q_0, then computation u enqueues its outputs meant for v into q_0. When computation v is ready, it dequeues all the inputs that it expects from u, from queue q_0. For this scheme to be successful, when v dequeues its inputs from q_0, the outputs of u must be at the front of the queue. The problem of determining an assignment of arcs (dependencies) to the least number of queues is simply the problem of finding an optimal queue layout of the dependency dag. Similarly, the flow of data in a system of parallel processors can be managed by a set of stacks, and determining an assignment of the arcs (dependencies) to the least number of stacks corresponds to finding an optimal stack layout of the dependency dag.

Another motivation for stack and queue layouts of dags has emerged from the study of the stacknumber of partially ordered sets (posets) initiated by Syslo [**9**] and Nowakowski and Parker [**7**] and the study of the queuenumber of posets initiated by Heath and Pemmaraju [**5**].

3. Results

3.1. Layouts of Specific Classes of dags. In this section, we focus on several classes of dags classified according to the structure of their underlying undirected graphs.

THEOREM 1. *Let $\mathcal{T}$ be the class of all dags whose underlying undirected graph is a tree. Then $SN(\mathcal{T}) = 1$ and $QN(\mathcal{T}) = 2$.*

An example of a directed tree with queuenumber equal to 2 is given in Figure 1 (a).

THEOREM 2. *Let $\mathcal{U}$ be the class of all dags whose underlying graph is unicyclic (contains exactly one cycle). Then $SN(\mathcal{U}) = QN(\mathcal{U}) = 2$.*

An example of a dag whose underlying undirected graph is unicyclic and has stacknumber and queuenumber equal to 2 is given in Figure 1 (b).

Results like those of Theorems 1 and 2 give information about the worst behaved dags in a class. It is also interesting to consider the best behaved dags—those with queuenumber equal to 1 or stacknumber equal to 1. In the *undirected* case, a graph has stacknumber equal to 1 if and only if it is outerplanar [**1**]. Moreover, outerplanarity can be tested in linear time [**10**]. Heath and Rosenberg [**6**] show that the set of *undirected* graphs with queuenumber equal to 1 has a geometric interpretation as a class of planar graphs called *arched leveled-planar graphs*. However, determining whether a graph is arched leveled-planar is NP-complete [**6**].

The dags with queuenumber equal to 1 form another special class of directed planar graphs which we call *arched leveled-planar dags* and define as follows. Place a Cartesian coordinate system on the plane and define the vertical line $\ell_i = \{(i, y) | y \in \mathit{Reals}\}$ where i is an integer. A dag $\vec{G}$ is *leveled-planar* if its vertex set V can be partitioned $V = V_1 \cup V_2 \cup \ldots \cup V_m$ so that (i) $\vec{G}$ has a planar embedding in which all vertices of V_i are on line ℓ_i, and (ii) each arc (u, v) is embedded as a straight line segment with $u \in V_i$ and $v \in V_{i+1}$ for some $i = 1, 2, \ldots m-1$. Such an embedding is called a *leveled-planar embedding.* A *level-i arch* is an arc (u, v) that satisfies: (i) both u and v lie on the line ℓ_i, (ii) node v is the top node on line ℓ_i, and (iii) there is no node on line ℓ_i below u that has a neighbor on line ℓ_{i+1}. We define an *arched leveled-planar dag* to be a leveled-planar dag augmented by any number of level-i arches for each i.

An arched leveled-planar embedding of a dag induces a natural ordering σ on the nodes, namely, as i takes the values $1, 2, \ldots m$, the vertices on line ℓ_i are listed from bottom to top. This ordering σ gives a 1-queue layout of an arched leveled-planar dag. The proof of the converse (that any dag with queuenumber 1 has an arched leveled-planar embedding) is similar to that of the analogous result for undirected graphs given in [**6**]. While the result for dags gives a nice geometric characterization of dags with queuenumber 1, it does not immediately yield a recognition algorithm. We conjecture that the class of dags with queuenumber equal to 1 can be recognized in polynomial time, in contrast with the undirected case. We hope to include such a recognition algorithm in the final version of this paper.

If we restrict attention to specific classes of dags, we obtain more tractable characterizations of those dags with queuenumber equal to 1. The next theorem gives a forbidden graph characterization of those dags whose underlying undirected graphs are trees and whose queuenumber is 1. First define a *leveling* for a dag whose underlying undirected graph is a tree. A *leveling* is a function *lev* that assigns an integer to each node of the dag, whose underlying undirected

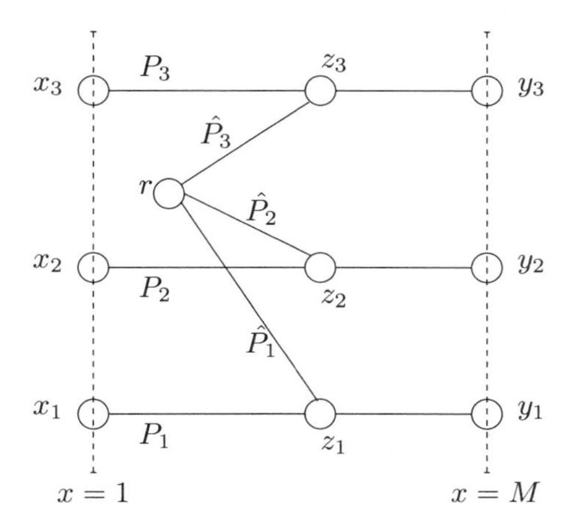

FIGURE 2. A typical dag in the forbidden set $\mathcal{F}$.

graph is a tree, so that $lev(u) = lev(v) + 1$ for each arc (u, v). Once a node is designated as the root of the tree, and its level fixed, the rest of the leveling is uniquely determined. Now define a set $\mathcal{F}$ of dags $\vec{F}$ that satisfy the following. Each dag $\vec{F}$ consists of two sets of three paths; $\{P_1, P_2, P_3\}$ and $\{\hat{P}_1, \hat{P}_2, \hat{P}_3\}$ (refer to Figure 2). There is a positive integer M and a leveling lev of $\vec{F}$ for which all nodes u of $\vec{F}$ satisfy $1 \leq lev(u) \leq M$. Furthermore, the paths P_i, with endpoints x_i and y_i, are node disjoint and have $lev(x_i) = 1$ and $lev(y_i) = M$ for $i = 1, 2, 3$. The paths $\hat{P}_i$ have endpoints r and z_i and share exactly one node r. The paths $\hat{P}_i$ and P_j share the node z_i if $i = j$ and are node disjoint otherwise. A typical dag in the set $\mathcal{F}$ is shown in Figure 2. Note that the dag in Figure 1 (a) is a smallest dag in the set $\mathcal{F}$.

THEOREM 3. *Let $\vec{T}$ be a dag whose underlying undirected graph is a tree. Then $\vec{T}$ is a 1-queue dag iff $\vec{T}$ has no (directed) subgraph in the set $\mathcal{F}$ described above.*

3.2. Stacks vs. Queues. The following theorem shows a contrast between the stacknumber and the queuenumber of classes of dags.

THEOREM 4. *There exists a class of outerplanar dags $\mathcal{C}$ such that $QN(\mathcal{C}) = \Omega(n)$ and $SN(\mathcal{C}) = 1$. There exists a class of planar dags $\mathcal{D}$ such that $SN(\mathcal{D}) = n$ and $QN(\mathcal{D}) = 2$.*

In the undirected case, there are classes of undirected graphs $\mathcal{G}$ for which the ratio $SN(\mathcal{G})/QN(\mathcal{G})$ is unbounded (see [**3**]) but it is not known whether there is a class of undirected graphs $\mathcal{G}$ for which the ratio $QN(\mathcal{G})/SN(\mathcal{G})$ is unbounded.

Yannakakis [**11**] shows that planar graphs have stacknumber at most 4, and Heath, Leighton, and Rosenberg [**3**] show that outerplanar graphs have queuenumber at most 2. Hence we obtain the following corollary of Theorem 4 which reveals a marked contrast in the stacknumber (queuenumber) of a dag and the stacknumber (queuenumber) of its underlying undirected graph.

COROLLARY 5. *If $\mathcal{D}$ is the class of outerplanar* dags *and $\mathcal{G}$ is the class of outerplanar* graphs, *then the ratio $\frac{QN(\mathcal{D})}{QN(\mathcal{G})}$ is unbounded. If $\mathcal{D}$ is the class of planar* dags *and $\mathcal{G}$ is the class of planar* graphs, *then the ratio $\frac{SN(\mathcal{D})}{SN(\mathcal{G})}$ is unbounded.*

While the above corollary shows that the queuenumber of a dag and the queuenumber of its underlying undirected graph may differ markedly, we do have an upper bound on the queuenumber of a dag in terms of the length of the longest directed path in the dag and the queuenumber of the underlying undirected graph.

THEOREM 6. *Let $\vec{G}$ be a dag, let $L(\vec{G})$ be the longest directed path in $\vec{G}$, and let G be the underlying undirected graph of $\vec{G}$. Then*

$$QN(\vec{G}) \leq 2 \cdot L(\vec{G}) \cdot QN(G).$$

3.3. NP-Completeness Results. For undirected graphs, the problems of finding an optimal queue layout and an optimal stack layout are known to be NP-complete. The next theorem shows that these problems are also NP-complete in the directed case. Define the decision problems DIRECTED STACKNUMBER and DIRECTED QUEUENUMBER as follows.

DIRECTED STACKNUMBER
INSTANCE: A dag $\vec{G} = (V, \vec{E})$.
QUESTION: Can $\vec{G}$ be laid out in 9 stacks?

DIRECTED QUEUENUMBER
INSTANCE: A dag $\vec{G} = (V, \vec{E})$.
QUESTION: Can $\vec{G}$ be laid out in 7 queues?

THEOREM 7. *DIRECTED STACKNUMBER and DIRECTED QUEUENUMBER are NP-complete.*

4. Open Problems

We mention a sampling of open problems about stack and queue layouts of dags.

- Find a polynomial time algorithm for recognizing 1-queue dags. The existence of such an algorithm would be in contrast with a result of

Heath and Rosenberg [6] and would also be in contrast with the NP-completeness results in this paper.

- Give a forbidden graph characterization of the class of 1-queue dags.
- Find the stacknumber of dags whose underlying undirected graphs are outerplanar.
- Find an upper bound on the stacknumber of a dag in terms of the stacknumber of its underlying undirected graph and the length of the longest directed path in it (see Theorem 6).

References

1. Frank Bernhart and Paul C. Kainen, *The book thickness of a graph*, Journal of Combinatorial Theory **27** (1979), 320–331.
2. Fan R. K. Chung, Frank Thomson Leighton, and Arnold L. Rosenberg, *Embedding graphs in books: a layout problem with applications to VLSI design*, SIAM Journal on Algebraic and Discrete Methods **8** (1987), 33–58.
3. Lenwood S. Heath, Frank Thomson Leighton, and Arnold L. Rosenberg, *Comparing queues and stacks as mechanisms for laying out graphs*, SIAM Journal on Discrete Mathematics **5** (1992), 398–412.
4. Lenwood S. Heath and Sriram V. Pemmaraju, *Queue layouts and matrix covers*, Submitted, 1992.
5. ———, *Stack and queue layouts of posets*, Submitted, 1992.
6. Lenwood S. Heath and Arnold L. Rosenberg, *Laying out graphs using queues*, To appear, SIAM Journal on Computing, 1992.
7. Richard Nowakowski and Andrew Parker, *Ordered sets, pagenumbers and planarity*, Order **6** (1989), 209–218.
8. Christos H. Papadimitriou and Mihalis Yannakakis, *Towards an architecture-independent analysis of parallel algorithms*, Proceedings of the 20th Annual ACM Symposium on Theory of Computing, 1988, pp. 510–513.
9. Maciej M. Sysło, *Bounds to the page number of partially ordered sets*, Tech. Report 227/1989, Technische Universität Berlin, 1989.
10. Maciej M. Sysło and Masao Iri, *Efficient outerplanarity testing*, Fundamenta Informaticae (1979), 261–275.
11. Mihalis Yannakakis, *Four pages are necessary and sufficient for planar graphs*, Proceedings of the 18th Annual ACM Symposium on Theory of Computing, 1986, pp. 104–108.

Department of Computer Science, Virginia Polytechnic Institute and State University, Blacksburg, VA 24061

E-mail address: heath@heath.cs.vt.edu

Department of Computer Science, Virginia Polytechnic Institute and State University, Blacksburg, VA 24061

E-mail address: sriram@cstheory.cs.vt.edu

Department of Mathematics, Wellesley College, Wellesley, MA 02181

E-mail address: atrenk@lucy.wellesley.edu

DIMACS Series in Discrete Mathematics
and Theoretical Computer Science
Volume 9, 1993

Enumeration of Degree Restricted Maps on the Sphere

E. A. BENDER AND E. R. CANFIELD

September 23, 1992

ABSTRACT. We find the generating function for n-edged rooted maps on the sphere, all of whose face degrees belong to a prescribed set D.

Recall that a *rooted map* is a connected graph, drawn on the plane, with a distinguished edge, the *root edge*, having a specified direction and a specified side. A homemorphism of the plane preserving all graph structure – incidences, root edge, root edge orientation and side – is a combinatorial equivalence, and we wish to count maps up to such equivalence.

To put the present work into perspective, let us recall the original contribution of Tutte [**2**] in this subject. We form the generating function

$$M(x,y) = \sum_{\mathcal{M}} x^{e(\mathcal{M})} y^{r(\mathcal{M})},$$

in which the summation is over all rooted maps $\mathcal{M}$, $e(\mathcal{M})$ and $r(\mathcal{M})$ being the number of edges and the root face degree of $\mathcal{M}$. Although a_n, the number of planar rooted maps with n edges, is the coefficient of x^n in the single-variable generating function $M(x,1)$, it is necessary to consider this more involved function $M(x,y)$ first. A rooted planar map is either a single point, or it has a root edge which is either a bridge or not; this analysis leads to the generating function equation

$$M = 1 + xy^2M^2 + x(yM_0 + (y^2+y)M_1 + (y^3+y^2+y)M_2 + \cdots),$$

1991 *Mathematics Subject Classification.* Primary 05A15, 05A16, 05C10.

The research of the second author was supported in part by the National Science Foundation.

This is an extended abstract of a paper submitted for publication elsewhere.

in which $M = M(x,y)$ and $M_k = M_k(x)$ is the single-variable generating function for maps, by number of edges, over those maps whose rootface degree is k. After some algebra and completing the square, we find that

$$\left(2xy^2(1-y)M-(1-y+xy^2)\right)^2 = (1-y+xy^2)^2-x(1-y)^2y^2-x^2y^3(1-y)M(x,1),$$

which says that the power-series in parentheses on the left, although containing infinitely many terms with the variable y, yields a polynomial $p(y)$ of degree 4 in y, with coefficients which are formal power series in x, after being squared. It follows that the discriminant of $p(y)$ must be zero, since there is a power series $Y = Y(x)$ in x causing both $p(Y)$ and $p'(Y)$ to vanish. This tells us that

$$27x^2M(x,1)^2 - (18x-1)M(x,1) + (16x-1) = 0$$

and thence

$$a_n = \frac{2(2n)!3^n}{n!(n+2)!},$$

the classical formula of Tutte.

Now let us consider a second example: enumerate planar rooted maps in which all non-root faces are required to have degree 2, 3, or 4. Proceding as before,

$$M = 1+xy^2M^2+x(yM_1+(y^2+y)M_2+(y^3+y^2+y)M_3+(y^4+y^3+y^2)M_4+\cdots),$$

leading to the conclusion that the square of $2xy^4M - y^2 + x(1+y+y^2)$ is a polynomial in y of degree 6. However, when we use the "discriminant trick" which worked for evaluation of a_n, we obtain only a relation between two unknown power-series, not a complete solution. Instead, we consider the following equation, (see [**1**]),

$$(1) \quad \left(2xy^4M - y^2 + x(1+y+y^2)\right)\left(1 + 2R_1y + (6R_1^2 + 2R_2)y^2 + \cdots\right) = x + Q_1y + Q_2y^2.$$

The second factor on the left side is the power series expansion for $(1-4R_1y-4R_2y^2)^{-1/2}$. When we compare the coefficients of y^0 in (1), we find $x = x$; comparing y^1 and y^2 we find Q_1 and Q_2 expressed, respectively, as functions of R_1 and R_2; on comparing the coefficients of y^3 and y^4 (using the fact that $M_0 = 1$) we obtain a pair of equations involving R_1 and R_2; later coefficients of $y^5, y^6, \ldots$ yield expressions for $M_1, M_2, \ldots$ in terms of R_1 and R_2. In short, the relations expressed by equation (1) lead to a parametric solution for M, and in particular (1) may be rewritten

$$\left(2xy^4M - y^2 + x(1+y+y^2)\right)^2 = (x + Q_1y + Q_2y^2)^2(1 - 4R_1y - 4R_2y^2),$$

with Q_i and R_i determined as described above. This technique turns out to be of general applicability, and we have the following.

THEOREM. *Let $D \subseteq \{1, 2, \dots\}$ be a set, and $m(n)$ be the number of rooted maps on the sphere all of whose face degrees belong to the set D. Then,*

$$(1 + \sum_{n \geq 1} m(n) x^n) x = M_2(x),$$

and

$$M_2'(x) = \frac{(R_1^2 + R_2)(9R_1^2 + R_2)}{x^2},$$

where R_1 and R_2 are the formal power series solutions to the pair of equations

$$R_1 = \frac{x}{2} \sum_{i \in D} [y^{i-1}](1 - 4R_1 y - 4R_2 y^2)^{-1/2}$$

$$R_2 = x + \frac{x}{2} \sum_{i \in D} [y^i](1 - 4R_1 y - 4R_2 y^2)^{-1/2} - 3R_1^2.$$

Some applications of this Theorem are (1) to identify circumstances in which M(x,y) is algebraic; (2) to recover Tutte's formula

$$\frac{2(nd)!}{n!(nd - n + 2)!} \binom{2d-1}{d}^n$$

for the number of $2d$-regular, nd-edged rooted planar maps; (3) to recover Y. Liu's (recently and independently found) formula for the number of rooted planar eulerian maps

$$\frac{3}{2(n+1)(n+2)} \binom{2n}{n} 2^n;$$

and (4) for the case when $D \subseteq \{2, 4, 6, \dots\}$ to give an asymptotic formula for $m(n)$ of the form

$$m(n) \sim c_D n^{-5/2} \gamma_D^n, \quad n \to \infty,$$

in which the dependence of c_D and γ_D on the set D is explicitly known.

What we have not been able to do is (1) give an asymptotic formula for the general case of D containing some odd integers; or (2) explain the formula

$$M_2'(x) = \left(\frac{R_2(x)}{x}\right)^2,$$

with

$$R_2 = x + \frac{x}{2} \sum_{2i \in D} \binom{2i}{i} R_2^i$$

which occurs when D contains even numbers only. The latter simplification comes about because in this case $R_1 = 0$; one can give a combinatorial interpretation to the result, but not yet a combinatorial explanation.

References

1. W. G. Brown, *On the existence of square roots in certain rings of power series*, Math. Annalen **158** (1965), 82–89.
2. W. T. Tutte, *A census of planar maps*, Canadian J. of Math. **15** (1963), 249–271.

Department of Mathematics, UCSD, La Jolla, California 92093
E-mail address: ebender@euclid.ucsd.edu

Department of Computer Science, University of Georgia, Athens, Georgia 30602
E-mail address: erc@pollux.cs.uga.edu

DIMACS Series in Discrete Mathematics
and Theoretical Computer Science
Volume 9, 1993

A Generalisation of MacLane's Theorem to 3-Graphs

C. PAUL BONNINGTON AND CHARLES H. C. LITTLE

July 31, 1992

ABSTRACT. A 3-graph is a cubic graph endowed with a proper edge colouring in three colours. A special type of 3-graph, called a gem, can be used to provide a natural combinatorial method to model graph imbeddings. For example, a spherical gem models an imbedding of a graph on the sphere. Two gems are said to be congruent if they model imbeddings of the same graph. In this paper we generalise the concepts of sphericality and congruence to 3-graphs. A theorem is given which shows that 3-graphs congruent to spherical 3-graphs are characterised by the existence of a set of circuits which interact with the different coloured edges in a special way. This theorem is applied to the case of gems to obtain MacLane's famous test of planarity [**8**].

1. Introduction

Throughout this paper, the sum of sets is defined as their symmetric difference. The graphs we consider lack loops, unless we indicate otherwise, but may have multiple edges. This paper is concerned only with finite graphs, those graphs G in which the vertex set VG and the edge set EG are both finite. Two distinct edges of a graph are said to be *adjacent* if they are incident on a common vertex. We write $c(G)$ for the number of components in a graph G. If $T \subseteq EG$, then we write $G[T]$ for the subgraph of G whose edge set is T and whose vertex set is the set of all vertices of G incident with at least one edge of T. If $S \subseteq VG$, then we define ∂S to be the set of edges which join a vertex of S to a vertex of $VG - S$; hence $\partial S = \partial(VG - S)$.

A *path* P joining two vertices, a and b, in the same component of G is the edge set of a minimal connected subgraph of G containing a and b. If x and y

---

1991 *Mathematics Subject Classification.* Primary 05C10, 05C75.

This paper is in final form and no version of it will be submitted for publication elsewhere.

are vertices incident on edges of a path P, then we denote by $P[x, y]$ the path in $G[P]$ joining x and y. A *circuit* in G is the edge set of a non-empty connected subgraph in which each vertex has degree 2. The *length* of a path or circuit is its cardinality. If C is a circuit or a path, the elements of $VC = VG[C]$ are sometimes referred to as vertices of C. Suppose C is a circuit of length greater than 1, and let v be a vertex of C. Let a and b be the edges of C incident on v. Then $C - \{a, b\}$ is a path, which we denote by C_v.

The *cycle space* of G is the vector space (over the field of residue classes modulo 2) spanned by the set of circuits of G. We denote it by $\mathcal{Z}(G)$, and its elements are *cycles* .

Let K be a graph. A *proper edge colouring of* K is a colouring of the edges so that adjacent edges receive distinct colours. A 3-*graph* is defined as an ordered triple $(K, \mathcal{P}, \mathcal{O})$ where K is a cubic graph endowed with a proper edge colouring $\mathcal{P}$ in three colours and $\mathcal{O}$ is a ordering of the three colours. We shall assume throughout that the three colours are blue, yellow and red. We write $K = (K, \mathcal{P}, \mathcal{O})$ when no ambiguity results. If H is a set of edges in K or a subgraph of K, then we write $\beta(H)$ and $\rho(H)$ for the set of blue and red edges respectively in H.

The set obtained from EK by deletion of the edges of a specified colour is the union of a set of disjoint circuits, called *bigons.* Thus bigons are of three types: red-yellow, red-blue and blue-yellow. We denote the sets of red-yellow, red-blue and blue-yellow bigons by $\mathcal{B}(K)$, $\mathcal{Y}(K)$ and $\mathcal{R}(K)$ respectively. The total number of bigons is $r(K) = |\mathcal{B}(K)| + |\mathcal{Y}(K)| + |\mathcal{R}(K)|$.

EXAMPLE 1. Consider the 3-graph K in Figure 1. Evidently

$$\begin{aligned} \mathcal{B}(K) &= \{\{a_1, c_2, a_3, c_4, a_4, c_3, a_2, c_1\}\}, \\ \mathcal{Y}(K) &= \{\{a_1, b_2, a_4, b_1\}, \{a_2, b_3, a_3, b_4\}\}, \text{ and} \\ \mathcal{R}(K) &= \{\{c_1, b_1, c_3, b_4, c_4, b_2, c_2, b_3\}\}. \end{aligned}$$

Hence $r(K) = 4$.

Following Lins [**4, 5**], we define a *gem* to be a 3-graph in which the red-blue bigons are quadrilaterals (circuits of length 4). For example, the 3-graph in Example 1 is a gem. We often say that red-blue bigons in a gem are red-blue *bisquares.*

A 2-cell imbedding of a graph G, which may have loops, in a closed surface $\mathcal{S}$ can be modelled by means of a gem in the following way (see [**1, 4, 5, 7**]). First construct the barycentric subdivision Δ of the imbedding of G, and colour each vertex of Δ with blue, yellow or red according to whether it represents a vertex, edge or face of the imbedding. Each edge of Δ then joins vertices of distinct colours, and may be coloured with the third colour. Let K be the dual graph of Δ, each edge of K being coloured with the colour of the corresponding edge of Δ. Then each red-blue bigon of the 3-graph K is a quadrilateral, so that K is a gem. (See Figure 2. In this figure, the vertices of G are the solid circles and the

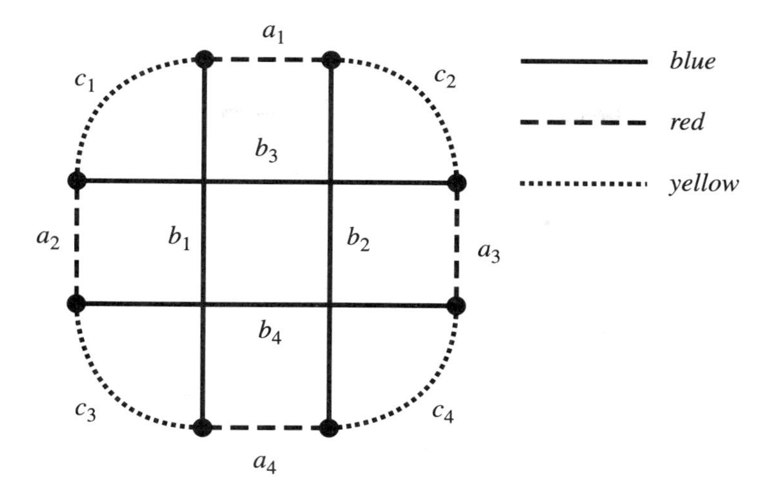

FIGURE 1.

edges are the thin solid lines joining such circles. All the circles are vertices of Δ. The edges of Δ are thin solid line segments and the thin dashed lines. The edges of K are thicker and coloured as indicated in the figure. The vertices of K should be self-evident.)

This construction can be reversed. Given K, we first contract each red-yellow bigon to a single vertex. Each red-blue bigon then becomes a digon whose edges are both blue. The identification of the two edges in each of these digons yields G. Thus there is a 1:1 correspondence between gems and 2-cell imbeddings of graphs in closed surfaces. Also, if $S \subseteq EK$ then the blue edges in S appear in G. The set T of such edges of G is also said to *correspond* to S, and vice versa, but this correspondence is not 1:1. In general, T corresponds to several subsets of EK. We also say that each of these subsets *represents* T.

If G is obtained from a gem K in this way, we say that G *underlies* K. We also say that K *represents* the imbedding of G.

The vertices of G are in 1:1 correspondence with the red-yellow bigons of K, the edges of G with the red-blue bigons of K, and the faces of the imbedding of G with the blue-yellow bigons of K. Thus we have $r(K) = |VG| - |EG| + |FG|$, where FG is the set of faces of the imbedding of G. The Euler characteristic $\chi(\mathcal{S})$ of $\mathcal{S}$ is therefore

$$|VG| - |EG| + |FG| = r(K) - 2|EG| = r(K) - \frac{|VK|}{2}$$

since $|VK| = 4|EG|$. In general, if K is a 3-graph then we define the *Euler characteristic* of K to be

$$\chi(K) = r(K)J - J\frac{|VK|}{2}.$$

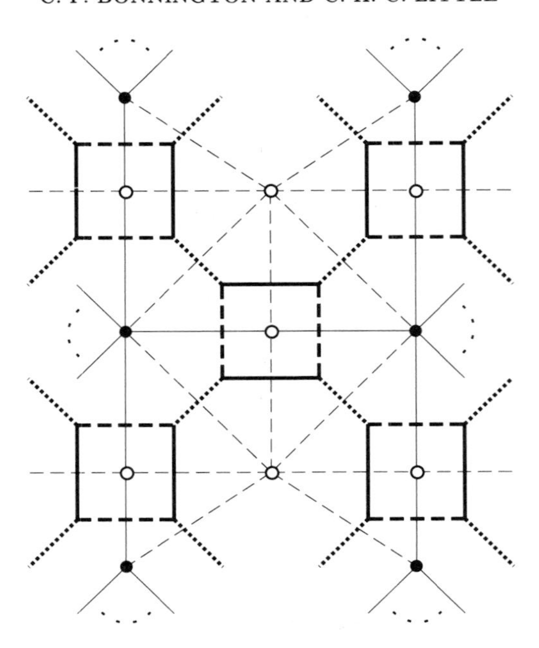

FIGURE 2.

In the special case where $\chi(K) = 2c(K)$, we say that K is *spherical.* The motivation for this definition comes from the fact that if a connected graph G underlies a spherical gem, then such a gem models a planar embedding for G.

Gems appeared first in the doctoral dissertation of Robertson [**9**] and subsequently in work of Ferri and Gagliardi [**2**]. The correspondence between gems and imbeddings was developed by Lins in [**4, 5**], though his account was not expressed in terms of the barycentric subdivision of the imbedding. We shall see how this non-topological axiomatic definition for a graph imbedding means that no topological apparatus need be brought into play when proving theorems in topological graph theory.

A graph G may underlie many gems – indeed, each such gem models an imbedding of G in some surface. This observation motivates the following definition. Let K and L be two 3-graphs. Suppose there exist bijections θ, φ, σ between $\mathcal{B}(L)$ and $\mathcal{B}(K)$, $\beta(L)$ and $\beta(K)$, and $\rho(L)$ and $\rho(K)$ respectively. Furthermore, suppose that

(i) for any red-yellow bigon B in L and any red edge $a \in B$ we have $\sigma(a) \in \theta(B)$, and

(ii) for any blue edge b adjacent to a red edge a we have $\varphi(b)$ adjacent to $\sigma(a)$.

Then K and L are *congruent.* Thus two gems are congruent if and only if a graph underlies them both. Moreover, if K and L are congruent then by condition (ii) we have a bijection between the red-blue bigons of K and the red-blue bigons of L. Evidently, congruence is an equivalence relation.

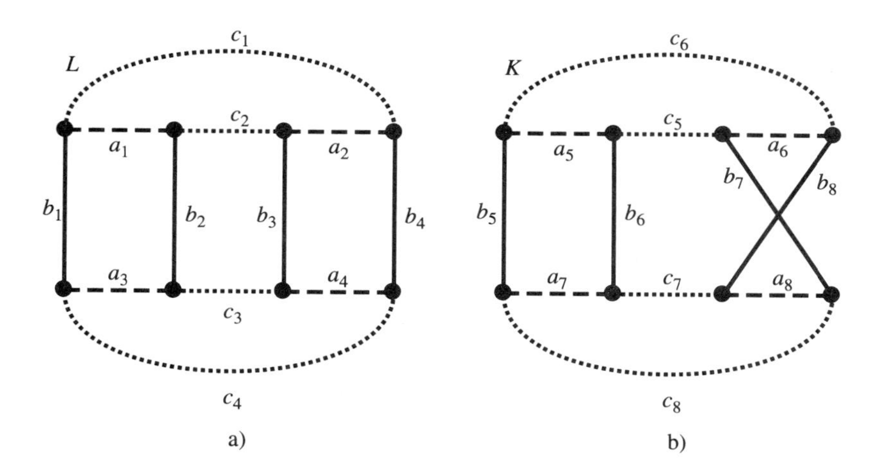

FIGURE 3.

EXAMPLE 2. Consider the gems L and K in Figure 3. Let B_1, B_2, B_3 and B_4 denote the red-yellow bigons $\{a_1, c_1, a_2, c_2\}$, $\{a_3, c_3, a_4, c_4\}$, $\{a_5, c_5, a_6, c_6\}$, and $\{a_7, c_7, a_8, c_8\}$ respectively. Define

$$\begin{aligned} \theta(B_i) &= B_{i+2} && \text{for } 1 \leq i \leq 2, \\ \varphi(bi) &= b_{i+4} && \text{for } 1 \leq i \leq 4, \\ \sigma(a_i) &= a_{i+4} && \text{for } 1 \leq i \leq 4. \end{aligned}$$

Now, observe that

(i) for any red-yellow bigon B_j in L and any red edge $a_i \in B_j$ we have $\sigma(a_i) = a_{i+4} \in \theta(B_j) = B_{j+2}$, and

(ii) for any blue edge b_i adjacent to a red edge a_j we have $\varphi(b_i) = b_{i+4}$ adjacent to $\sigma(a_j) = a_{i+4}$.

Hence we conclude that K and L are congruent. As expected, a graph G underlies them both and this is the connected graph with two vertices and two edges and no loops. L models an imbedding of G on the sphere and K models an imbedding of G on the projective plane.

If K and L are congruent 3-graphs and $E = \{e_1, e_2, \ldots, e_n\}$ is a set of red (blue) edges in L, then for conciseness we usually write E for $\sigma(E)$ $(\varphi(E))$ and e_i for $\sigma(e_i)$ $(\varphi(e_i))$ when no ambiguity results. MacLane [**8**], in an attempt to make a partial separation between graph theory and topology, proved that a given graph can be imbedded in the sphere if and only if it has a certain combinatorial property. One can translate MacLane's theorem into the language of gems to characterise which gems are congruent to spherical gems. However, in this paper we work in the more general setting of 3-graphs. Our main theorem gives the necessary and sufficient conditions for a 3-graph to be congruent to a spherical 3-graph. The paper is divided into six sections. In Section 2 we state our main theorem. It characterises 3-graphs that are congruent to spherical 3-

graphs by the existence of a special family of "semicycles" – a special type of circuit that was originally studied in the setting of permutation pairs by Stahl in [**10**]. Section 3 gives an introduction to the theory of dipoles in 3-graphs – a concept first developed in the topological setting by Ferri and Gagliardi in [**2**], but later explored in the combinatorial setting in [**1, 11**]. Sections 4 and 5 present the proof of our main theorem in both directions. In the final section, the topological graph theory implications of this theorem are discovered by specialising it to the case of gems to obtain MacLane's theorem.

2. Semicycles

Suppose C is a circuit containing at least one blue edge in a 3-graph K. We write $N(C)$ for the set of red-yellow bigons that contain an edge in C. Hence $N(C) \subseteq \mathcal{B}(K)$. We say that $N(C)$ is the *necklace* of C. The elements of its necklace are the *beads* of C. The *poles* of a bead B (with respect to C) are the vertices of B incident with a blue edge of C. If each bead has just two poles, then C is a *semicycle.*

The concept of a semicycle was introduced by Stahl [**10**] in the setting of permutation pairs, but the motivation for it is best explained by considering the case where K is a gem. As we indicated earlier, K then corresponds to a 2-cell embedding, in a closed surface $\mathcal{S}$, of a graph G. Under this interpretation, the beads of a semicycle C of K correspond to vertices of G. The requirements that C should be connected and have a blue edge, and that each bead should have just two poles, reveal that C corresponds to a circuit D of G or a path of length 1.

A *red-refined* 3-graph is one where all blue-yellow bigons are semicycles. A set $\mathcal{S}$ of semicycles in a 3-graph K is said to be a *semicycle cover* if the set of intersections of members of $\mathcal{S}$ with $\beta(K)$ is a partition of $\beta(K)$. Hence $\mathcal{R}(K)$ is a semicycle cover if K is red-refined. If $\mathcal{B}(K) \cup \mathcal{Y}(K) \cup \mathcal{S}$ spans the cycle space $\mathcal{Z}(K)$ of K we say that $\mathcal{S}$ is a *spanning semicycle cover.*

EXAMPLE 3. Consider the gem K of Figure 4. Let $C_1 = \{b_1, c_1, a_3, c_2, b_4, c_6\}$, $C_2 = \{b_3, a_3, c_2, a_4, b_6, c_5\}$ and $C_3 = \{b_5, c_2, a_4, c_3, b_2, c_4\}$. One can easily check that $\mathcal{B}(K) \cup \mathcal{Y}(K) \cup \{C_1, C_2, C_3\}$ spans the cycle space of K. Therefore $\mathcal{S}J = J\{C_1, C_2, C_3\}$ is a spanning semicycle cover.

We are now in a position to state our main theorem. Its proof follows from Theorems 3 and 4 which are given in Sections 4 and 5.

THEOREM 1. *A 3-graph K is congruent to a spherical 3-graph if and only if there exists a spanning semicycle cover in K.*

3. Dipoles

Let v and w be a pair of adjacent vertices in a 3-graph K. Suppose that v and w are linked by just one edge b, which is blue. Following Ferri and Gagliardi [**2, 3**], we say that b is a *blue* 1-*dipole* if the red-yellow bigons A and B passing

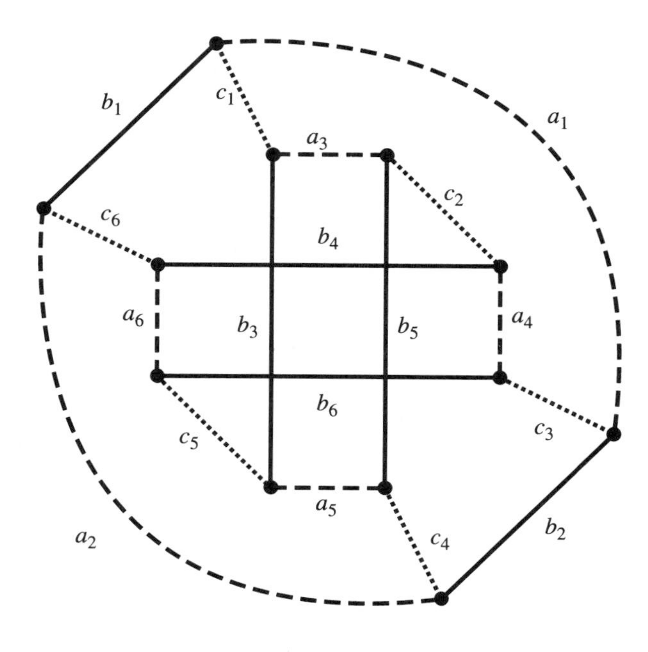

FIGURE 4.

through v and w respectively are distinct. Let c_1 and c_2 be the yellow edges incident on v and w respectively. Let a_1 and a_2 be the red edges incident on v and w respectively. Let v_1, v_2, w_1, w_2 be the vertices other than v and w incident on c_1, a_1, c_2, a_2 respectively. The *cancellation* of this blue 1-dipole b is the operation of deletion of the vertices v and w followed by the insertion of edges c and a linking v_1 to w_1 and v_2 to w_2 respectively. (See Figure 5.) We denote the resulting 3-graph by $K - [b]$. We observe that A and B have coalesced into one red-yellow bigon A'. The *creation* of a blue 1-dipole is the inverse operation. Similar definitions can be made for red and yellow 1-dipoles.

Now suppose that v and w are linked by two edges b and c coloured blue and yellow respectively. Following Ferri and Gagliardi [**2, 3**], we say that $\{b, c\}$ is a *blue-yellow* 2-*dipole* if the red edges a_1 and a_2 incident on v and w respectively are distinct. Let a_1 link v and v_1 and let a_2 link w and w_1. The *cancellation* of this blue-yellow 2-dipole is the operation of deletion of the vertices v and w followed by the insertion of an edge a linking v_1 to w_1. (See Figure 6.) We denote the resulting 3-graph by $K - [b, c]$. We observe that a_1 and a_2 have coalesced into one red edge a. The *creation* of a blue-yellow 2-dipole is the inverse operation. Similar definitions can be made for red-yellow and red-blue 2-dipoles.

We note that the red edge a_1 is a red 1-dipole in K and that the 3-graph $K - [a_1]$ is isomorphic to $K - [b, c]$. Hence cancellation or creation of a 2-dipole is in fact a special case of a 1-dipole cancellation or creation.

An intriguing result (see [**1, 3, 11**]) is that two 3-graphs, H and K, can be obtained from one another by a finite sequence of 1-dipole cancellations and

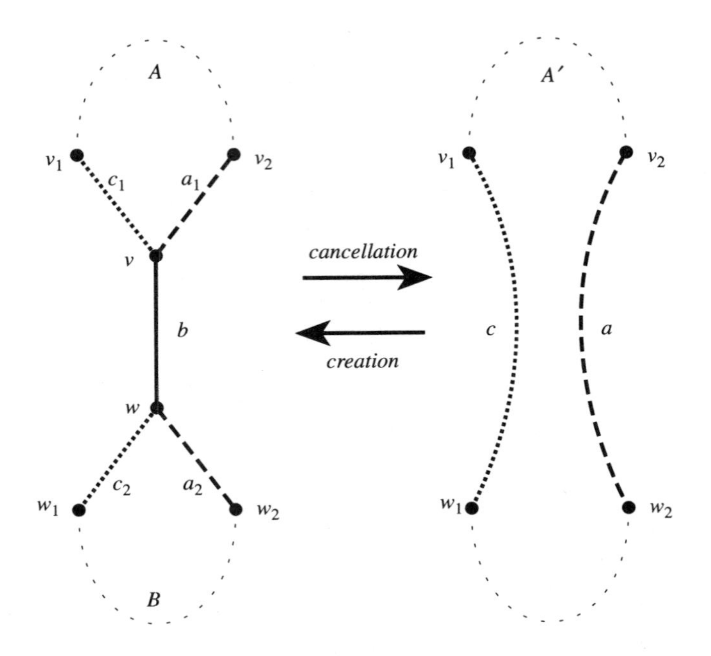

FIGURE 5.

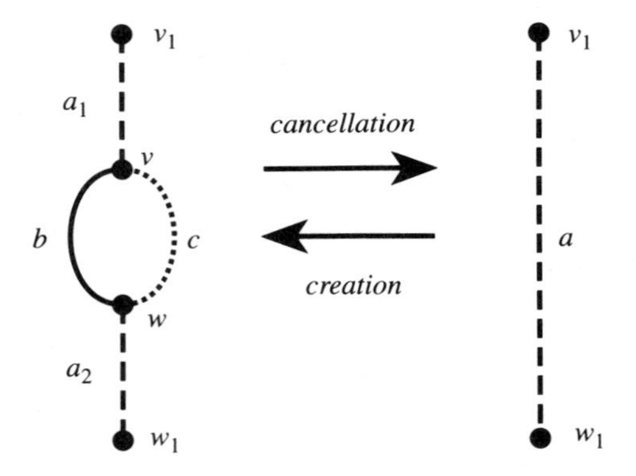

FIGURE 6.

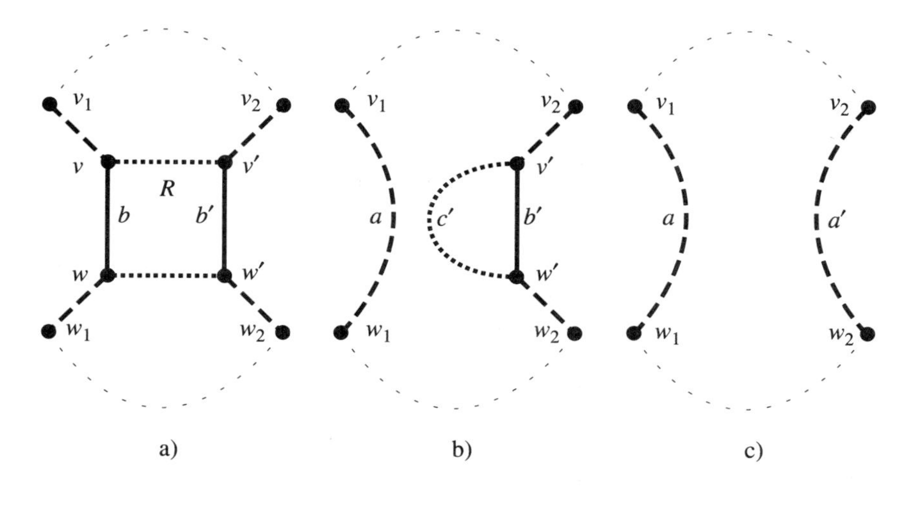

FIGURE 7.

creations if and only if H and K are both bipartite or both non-bipartite, and they have the same Euler characteristic. This result is a combinatorial analogue to the famous "classification of surfaces" theorem of Dehn and Heegaard.

Suppose R to be a blue-yellow bigon of length 4 in a 3-graph K. Label the edges and vertices incident on R as in Figure 7a. If b is a blue 1-dipole then let $K' = K - [b]$. (See Figure 7b.) Let c' denote the yellow edge of K' that joins v' and w'. If $\{b', c'\}$ is a blue-yellow 2-dipole in K' then let $K'' = K' - [b', c']$. (See Figure 7c.) We say that K'' is obtained from K by *cancellation* of the blue-yellow bigon R. Let a and a' denote the distinct red edges that join v_1 and w_1, and v_2 and w_2, respectively. The inverse operation is described as *splitting* a and a' to *create* the blue-yellow bigon R.

Let K be a 3-graph. Suppose there is a blue-yellow bigon R in K that is not a semicycle. Then there is a red-yellow bigon B such that $B \cap R$ contains two yellow edges c and c'. Hence $B - \{c, c'\}$ is the disjoint union of two paths P_1 and P_2 each of which contains a red edge. Let a be a red edge in P_1 and let a' be a red edge in P_2. Split a and a' to create the blue-yellow bigon R' and let K' denote the resulting graph. The red-yellow bigon B has now become two red-yellow bigons A and A' in K'. Furthermore R is a blue-yellow bigon in K' that s A in fewer edges than R meets B. A similar statement holds for A'. We also note that R' is a semicycle in K', and that any blue-yellow bigons in K' that are semicycles in K are semicycles in K'. Proceeding inductively we obtain a 3-graph such that R meets each red-yellow bigon in at most one edge, and hence is a semicycle. We conclude that it is possible to construct a 3-graph L from K by blue 1-dipole and blue-yellow 2-dipole creations such that all blue-yellow bigons in L are semicycles. Hence L is a red-refined 3-graph. We say that L is a *red-refinement* of K.

EXAMPLE 4. Consider the gem K in Figure 8a. Figure 8b illustrates a red-

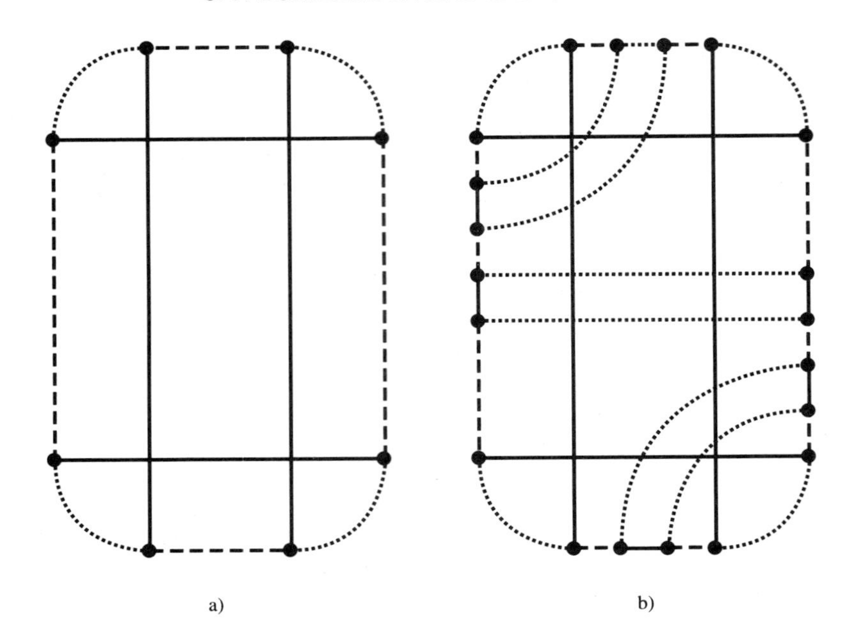

FIGURE 8.

refinement of K.

4. Boundary Covers

In this section we prove our main theorem in one direction. That is, any 3-graph congruent to a spherical 3-graph has a spanning semicycle cover. We prepare for its proof with some preliminary definitions and lemmas. The *boundary space* of a 3-graph K is the subspace of $\mathcal{Z}(K)$ spanned by the set of bigons of K. A semicycle cover $\mathcal{S}$ is a boundary cover if $\mathcal{B}(K) \cup \mathcal{Y}(K) \cup \mathcal{S}$ spans the boundary space of K. In [**6**], Little shows that the cycle space and the boundary space are equal in spherical 3-graphs. Hence boundary covers are also spanning semicycle covers in spherical 3-graphs.

LEMMA 1. *Let C and D be two cycles in a 3-graph K such that $\beta(C) = \beta(D)$. Then $C = D + \bigcup \mathcal{B}$ for some set $\mathcal{B}$ of red-yellow bigons in $N(C) = N(D)$.*

PROOF. $C + D$ is a cycle A which does not meet ∂VB for any bead B of C. Hence $A \cap B \in \{\emptyset, B\}$. If $A \cap B = \emptyset$ then $C \cap BJ = JD \cap B$, and if $A \cap B = B$ then $C \cap B = (D + B) \cap B$. Since $\beta(C) = \beta(D)$, it follows that $C = D + \bigcup \mathcal{B}$, where $\mathcal{B}$ is the set of beads B of C for which $A \cap B = B$. □

LEMMA 2. *Let K be a 3-graph with blue 1-dipole b and let $L = K - [b]$. There exists a boundary cover in L if there exists a boundary cover in K.*

PROOF. The following uses the notation of Figure 5. Let $\mathcal{S}$ denote a boundary cover of K. Let C_2 denote the semicycle in $\mathcal{S}$ that contains b. Suppose there exists a semicycle $C \in \mathcal{S}$ which contains c_1. Then clearly $C_1 = C + A$ is a semicycle such that $c_1 \notin C_1$, and $c_2 \in C_1$ if and only if $c_2 \in C$. Furthermore $\mathcal{S}_1 = (\mathcal{S} - \{C\}) \cup \{C_1\}$ is a boundary cover of K. A similar argument may be applied to c_2. Therefore we assume that no semicycle in $\mathcal{S}$ contains c_1 or c_2. In particular, C_2 includes $\{a_1, b, a_2\}$.

Let $\mathcal{S}_2 = (\mathcal{S} - \{C_2\}) \cup \{C_3\}$ where $C_3 = (C_2 - \{a_1, b, a_2\}) \cup \{a\}$. Evidently the set of intersections of members of $\mathcal{S}_2$ with $\beta(L)$ is a partition of $\beta(L)$. We claim that $\mathcal{B}(L) \cup \mathcal{Y}(L) \cup \mathcal{S}_2$ spans the boundary space of L. It is sufficient to show that any blue-yellow bigon R of L is a sum of circuits in $\mathcal{B}(L) \cup \mathcal{Y}(L) \cup \mathcal{S}_2$. If $c \in R$ let $R_1 = (R - \{c\}) \cup \{c_1, b, c_2\}$; otherwise let $R_1 = R$. Then R_1 is a blue-yellow bigon in K, and hence $R_1 = \sum \mathcal{U}$ for some set $\mathcal{U}$ of circuits in $\mathcal{B}(K) \cup \mathcal{Y}(K) \cup \mathcal{S}$. We claim that $A \in \mathcal{U}$ if and only if $B \in \mathcal{U}$. However this is clear, for $c_2 \in R_1$ if and only if $c_1 \in R_1$ and no semicycle in $\mathcal{S}$ contains either c_1 or c_2. Let Y denote the red-blue bigon in K that includes $\{a_1, b, a_2\}$ and let Y' denote the red-blue bigon in L that contains a. Let $\mathcal{U}_1$ be the set obtained from $\mathcal{U}$ by replacing C_2 with C_3, A and B with A', and Y with Y' if necessary. Evidently $R = \sum \mathcal{U}_1$ and $\mathcal{U}_1 \subseteq \mathcal{B}(L) \cup \mathcal{Y}(L) \cup \mathcal{S}_2$, as required.

Clearly C_3 is a semicycle in L. If $\mathcal{S}_2$ is a set of semicycles, then we are done. Suppose there is a circuit D in $\mathcal{S}_2$ that is not a semicycle in L. Then A' has 4 poles with respect to D, for otherwise D is not a semicycle in K. Let v, w, x, y denote the 4 poles, where $A'_y[v, w] \subset D$ and $x \in A'_v[w, y]$. Thus $A'_w[x, y] \subseteq D$. Clearly $C_4 = D_w[v, y] \cup A'_w[v, y]$ and $C_5 = D_v[w, x] \cup A'_v[w, Jx]$ are semicycles in L. Since $\beta(D)$ is the disjoint union of $\beta(C_4)$ and $\beta(C_5)$, then by Lemma 1, $C_4 + C_5 = D + \cup\mathcal{B}$ for some set $\mathcal{B}$ of red-yellow bigons in L. We conclude that $\mathcal{S}_3 = (\mathcal{S}_2 - \{D\}) \cup \{C_4, JC_5\}$ is a set of circuits, with one more semicycle than $\mathcal{S}_2$, such that $\mathcal{B}(L) \cup \mathcal{Y}(L) \cup \mathcal{S}_3$ spans the boundary space of L. Moreover, the set of intersections of members of $\mathcal{S}_3$ with $\beta(L)$ is a partition of $\beta(L)$. Proceeding inductively we obtain a boundary cover of L. □

LEMMA 3. *Let K be a 3-graph with blue-yellow 2-dipole $\{b, c\}$ and let $L = K - [b, c]$. Then there exists a boundary cover in L if there exists a boundary cover in K.*

PROOF. The following uses the notation of Figure 6. Let $\mathcal{S}$ denote a boundary cover of K. Let A be the red-yellow bigon in K that includes $\{a_1, c, a_2\}$. Let C denote the semicycle in $\mathcal{S}$ that contains b. Suppose $C \neq \{b, c\}$. Then clearly $\{b, c\} = C + A$. Furthermore $\mathcal{S}_1 = (\mathcal{S} - \{C\}) \cup \{C + A\}$ is a boundary cover of K. Therefore we assume that the semicycle C in $\mathcal{S}$ that contains b is $\{b, Jc\}$.

Let $\mathcal{S}' = \mathcal{S} - \{C\}$. Suppose there exists a semicycle $C_1 \in \mathcal{S}'$ which contains c. Then clearly $C_2 = C_1 + A$ is a semicycle such that $c \notin C_2$. Furthermore $\mathcal{S}_1 = (\mathcal{S} - \{C_1\}) \cup \{C_2\}$ is a boundary cover of K. Therefore we assume that no semicycle in $\mathcal{S}'$ contains c. Hence $\mathcal{S}'$ is a set of semicycles in L.

Evidently the set of intersections of members of $\mathcal{S}$ with $\beta(L)$ is a partition of $\beta(L)$. We claim that $\mathcal{B}(L) \cup \mathcal{Y}(L) \cup \mathcal{S}'$ spans the boundary space of L, and hence

S' is a boundary cover of L. It is sufficient to show that any blue-yellow bigon R of L is a sum of circuits in $\mathcal{B}(L) \cup \mathcal{Y}(L) \cup \mathcal{S}'$. Clearly R is a blue-yellow bigon in K, and hence $R = \sum \mathcal{U}$ for some set $\mathcal{U}$ of circuits in $\mathcal{B}(K) \cup \mathcal{Y}(K) \cup \mathcal{S}$. Let Y denote the red-blue bigon in K that contains b. Since no semicycle in $\mathcal{S}'$ contains b or c, then evidently $b \notin \bigcup(\mathcal{U} - \{C, Y\})$ and $c \notin \bigcup(\mathcal{U} - \{A, C\})$. Therefore, the fact that $R \notin \{b, c\}$ implies that either $\{A, C, Y\} \subseteq \mathcal{U}$ or $\{A, C, Y\} \cap \mathcal{U} = \emptyset$.

Let A' and Y' denote the red-yellow and red-blue bigons respectively in L that contain a. If $\{A, C, Y\} \subseteq \mathcal{U}$, then let $\mathcal{U}' = (\mathcal{U} - \{A, C, Y\}) \cup \{A', Y'\}$; otherwise let $\mathcal{U}' = \mathcal{U}$. Evidently $\mathcal{R} = \sum \mathcal{U}'$ and $\mathcal{U}' \subseteq \mathcal{B}(L) \cup \mathcal{Y}(L) \cup S'$, as required. □

THEOREM 2. *Every 3-graph has a boundary cover.*

PROOF. Let L be a red-refinement of an arbitrary 3-graph K. Therefore $\mathcal{R}(L)$ is a set of semicycles in L. It is immediate that $\mathcal{R}(L)$ is a boundary cover of L since the boundary space is the space spanned by $\mathcal{B}(L) \cup \mathcal{Y}(L) \cup \mathcal{R}(L)$. Hence our theorem follows from Lemmas 2 and 3, since K is obtained from L by a finite sequence of blue 1-dipole and blue-yellow 2-dipole cancellations. □

Let G be the graph obtained from a 3-graph K by contracting the red-yellow bigons to single vertices. We say that G is the *red-yellow reduction* of K. If K and L are congruent 3-graphs then evidently the red-yellow reductions of K and L are the same graph.

LEMMA 4. *Let G be the red-yellow reduction of a 3-graph K. Let*

$$\mathcal{C} = (C_1, JC_2, \dots, C_n)$$

be a family of circuits in G and let $\mathcal{D} = \{D_1, D_2, \dots, D_n\}$ be a set of semicycles in K such that $\beta(D_i) = C_i$ for all i. Then $\mathcal{C}$ spans $\mathcal{Z}(G)$ if and only if $\mathcal{B}(K) \cup \mathcal{D}$ spans $\mathcal{Z}(K)$.

PROOF. Firstly, assume that $\mathcal{C}$ spans $\mathcal{Z}(G)$ and let D be a circuit in K. We may assume that D contains a blue edge for otherwise D is a red-yellow bigon. Evidently $\beta(D)$ is a cycle in G and therefore $\beta(D) = \sum \mathcal{U}$ for some subset $\mathcal{U}$ of $\mathcal{C}$. Let $\mathcal{V}$ be the set of semicycles in $\mathcal{D}$ that correspond to the circuits in $\mathcal{U}$. Then clearly $\beta(\sum \mathcal{V}) = \beta(D)$. By Lemma 1, $D = \sum \mathcal{V} + \sum \mathcal{B}$ for some set $\mathcal{B} \subseteq \mathcal{B}(K)$. We conclude that $\mathcal{B}(K) \cup \mathcal{D}$ spans $\mathcal{Z}(K)$.

Now, assume that $\mathcal{B}(K) \cup \mathcal{D}$ spans $\mathcal{Z}(K)$ and let C be a circuit in G. Let D be a semicycle in K such that $\beta(D) = C$. Then $D = \sum \mathcal{V}$ for some subset $\mathcal{V}$ of $\mathcal{B}(K) \cup \mathcal{D}$. Let $\mathcal{U}$ be the set of circuits in $\mathcal{C}$ that correspond to the semi-cycles in $\mathcal{V} \cap \mathcal{D}$. Evidently, $C = \beta(D) = \beta(\sum \mathcal{V}) = \beta(\sum(\mathcal{V} \cap \mathcal{D})) = \sum \mathcal{U}$. We conclude that $\mathcal{C}$ spans $\mathcal{Z}(G)$, as required. □

THEOREM 3. *If K is congruent to a spherical 3-graph then there exists a spanning semicycle cover in K.*

PROOF. Let L be a spherical 3-graph congruent to K, and let G be the red-yellow reduction of K and L. By Lemma 2, there exists a boundary cover $\mathcal{S}'$ of L. Since L is spherical, $\mathcal{S}'$ is a spanning semicycle cover of L. Let $\mathcal{Y}$ be the set of cycles in G that correspond to the red-blue bigons in L. Let $\mathcal{R}$ be the set of circuits in G that correspond to the semicycles in $\mathcal{S}'$. By Lemma 4, we conclude that $\mathcal{Y} \cup \mathcal{R}$ spans $\mathcal{Z}(G)$.

For each circuit $R \in \mathcal{R}$ choose a semicycle in K that represents R and let $\mathcal{S}$ be the set of all semicycles so chosen. The fact that $\mathcal{R}$ is a partition of EG establishes that the set of intersections of members of $\mathcal{S}$ with $\beta(K)$ is a partition of $\beta(K)$. Let D be a circuit in K. Since an even number of blue edges in D are incident on a given red-yellow bigon, then $\beta(D)$ is a cycle in G. Therefore $\beta(D) = \sum \mathcal{U}$ for some set $\mathcal{U}$ consisting of cycles in $\mathcal{Y} \cup \mathcal{R}$. Let $\mathcal{V}$ be the set of semicycles in $\mathcal{S}$ that represent the circuits in $\mathcal{U} \cup \mathcal{R}$, and let $\mathcal{W}$ be the set of red-blue bigons in K that represent the circuits in $\mathcal{U} \cap \mathcal{Y}$. Therefore $\beta(D) = \sum \mathcal{U} = \beta(\sum(\mathcal{V} \cup \mathcal{W}))$. Hence, by Lemma 1, $D = \sum(\mathcal{V} \cup \mathcal{W}) + \bigcup \mathcal{B}$, for some set $\mathcal{B}$ of red-yellow bigons in $N(D)$. Thus D is a sum of bigons in $\mathcal{B}(K) \cup \mathcal{Y}(K) \cup \mathcal{S}$. We conclude that $\mathcal{S}$ is a spanning semicycle cover in K, as required. □

5. Partial Congruence

We now generalise the concept of congruence that was given in the introduction. Let K and L be two 3-graphs. Suppose there exist a partition $\mathcal{Q}$ of $\mathcal{B}(L)$ and bijections θ, φ, σ between $\mathcal{Q}$ and $\mathcal{B}(K), \beta(L)$ and $\beta(K)$ and $\rho(L)$ and $\rho(K)$ respectively. Furthermore, suppose that

(i) for any cell $\mathcal{B}$ of $\mathcal{Q}$ and any red edge $a \in \cup\mathcal{B}$ we have $\sigma(a) \in \theta(B)$, and
(ii) for any blue edge b adjacent to a red edge a we have $\varphi(b)$ adjacent to $\sigma(a)$.

Then we say that L is *partially congruent* to K (with respect to the partition $\mathcal{Q}$.)

If K and L are partially congruent 3-graphs and $E = \{e_1, e_2, \ldots, e_n\}$ is a set of red (blue) edges in L, then for conciseness we usually write E for $\sigma(E)(\varphi(E))$ and e_i for $\sigma(e_i)(\varphi(e_i))$ when no ambiguity results.

Let L be partially congruent to a 3-graph K and let $\mathcal{S}$ be a set of cycles in K. If for each $C \in \mathcal{S}$ there exists a cycle D in L such that $\beta(D) = \beta(C)$, then we say that L is *faithful* to K (with respect to $\mathcal{S}$.) The cycles D and C are said to *correspond* to each other.

EXAMPLE 5. Consider the 3-graphs K and L in Figure 9. Let $B_1 = \{a_1, Jc_3\}$, $B_2 = \{a_4, c_4\}$ and $B_3 = \{a_2, a_3, c_1, c_2\}$. Let $\mathcal{Q}$ be the partition $\{\{B_1, B_2\}, \{B_3\}\}$ of $\mathcal{B}(L)$. ¿From the labelling of the edges of K, it is clear that L is partially congruent to K with respect to the partition $\mathcal{Q}$. Let C be the cycle $\{b_2, c_4, b_3, c_1\}$ in K. Since $|\partial VB_1 \cap \beta(C)| = 1$, then $\beta(C)$ is not the blue edge set of a cycle in L. Hence L is not faithful to K with respect to $\{C\}$.

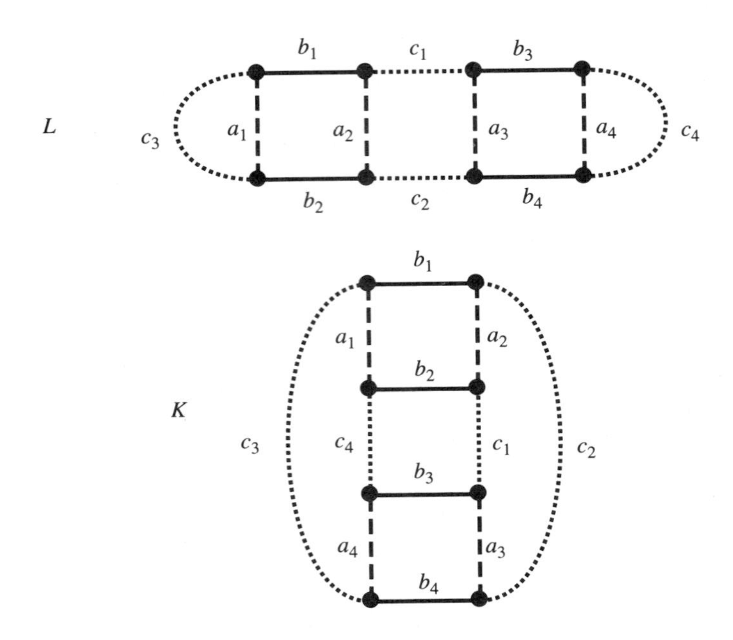

FIGURE 9.

LEMMA 5. *Let L be partially congruent to a 3-graph K. Let C be a cycle in L. Then $\beta(C)$ is the blue edge set of a cycle in K.*

PROOF. This follows from the fact that $|\beta(C) \cap \partial VB|$ is even for all $B \in \mathcal{B}(L)$, and therefore $|\beta(C) \cap \partial VB|$ is even for all $B \in \mathcal{B}(K)$. $\square$

LEMMA 6. *Let L be a 3-graph partially congruent to a 3-graph K with respect to a partition $\mathcal{Q}$. If L is faithful to K with respect to a spanning semicycle cover $\mathcal{S}$, then no two red-yellow bigons in a cell of $\mathcal{Q}$ belong to the same component of L.*

PROOF. We shall prove the contrapositive. Suppose B_1 and B_2 are distinct red-yellow bigons in a cell $\mathcal{B}$ of $\mathcal{Q}$ that belong to the same component L_1 of L. Let B denote the red-yellow bigon $\theta(B)$ in K. Since L_1 is connected, then there exists a path P, joining a vertex in VB_1 to a vertex in VB_2, such that $|PJ \cap \partial VB_1| = |P \cup \partial VB_2| = 1$ and $|P \cap \partial VB|$ is even for all $B \in \mathcal{B}(L) - \{B_1, JB_2\}$. Therefore $|P \cap \partial VB'|$ is even for all red-yellow bigons B' in K, since B_1 and B_2 both correspond to a single red-yellow bigon. Hence $\beta(P)$ is the blue edge set of a cycle in K.

Since $\mathcal{S}$ is a spanning semicycle cover, then $\beta(P)$ is the blue edge set of a cycle $\sum(\mathcal{Y} \cup \mathcal{U})$ for sets $\mathcal{Y} \subseteq \mathcal{Y}(K)$ and $\mathcal{U} \subseteq \mathcal{S}$. Let $\mathcal{R}$ be a set of cycles in L that correspond to the cycles in $\mathcal{U}$. Evidently, $\beta(P) = \beta(\sum(\mathcal{Y} \cup \mathcal{R}))$, and $\sum(\mathcal{Y}J \cup \mathcal{R})$ is a cycle in L. Hence $|P \cap \partial VB|$ is even for all $B \in \mathcal{B}(L)$, a contradiction to the fact that $|P \cap \partial VB_1|J = J|P \cap \partial VB_2| = 1$. $\square$

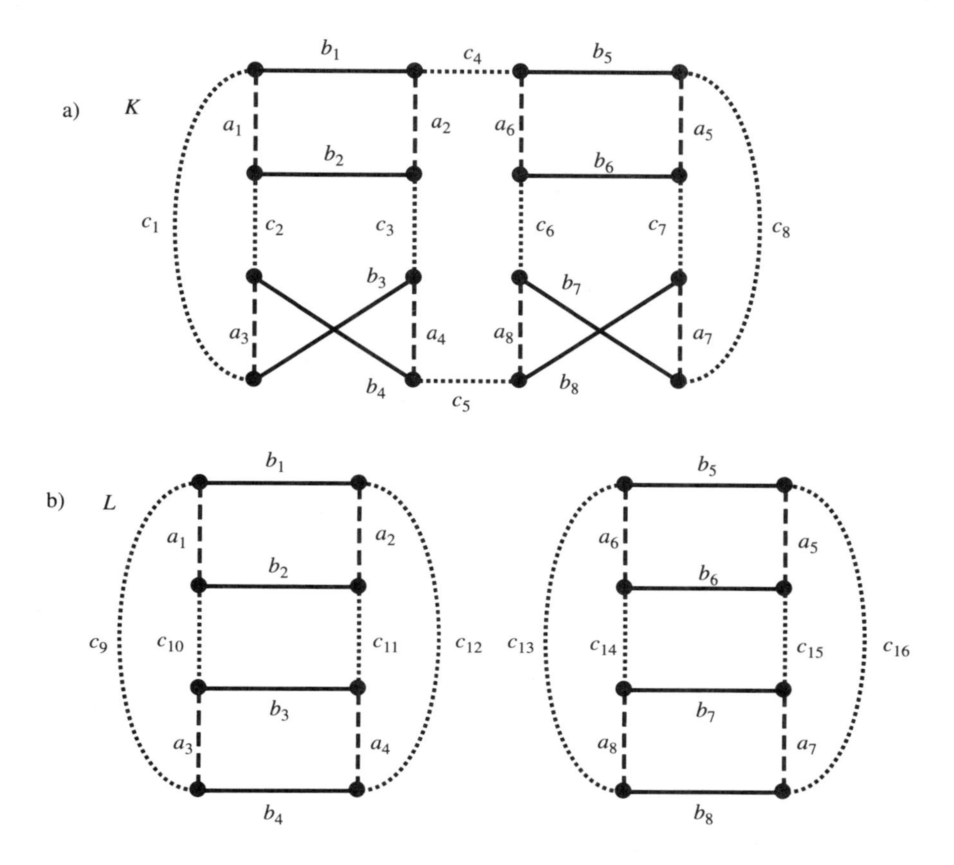

FIGURE 10.

EXAMPLE 6. Consider the 3-graphs K and L in Figures 10a and b respectively. Let

$$\begin{aligned} C_1 &= \{b_5, c_8, a_7, b_8, a_8, c_6, a_6\}, \\ C_2 &= \{b_6, c_7, a_7, b_7, c_6\}, \\ C_3 &= \{b_1, a_2, c_3, a_4, b_4, c_2, a_1\} \text{ and} \\ C_4 &= \{b_2, c_3, b_3, a_3, c_2\}. \end{aligned}$$

Let $\mathcal{S} = \{C_1, C_2, C_3, C_4\}$. One can easily check that $\mathcal{S}$ is a spanning semicycle cover in K. Let B_1, B_2, B_3 and B_4 be the red-yellow bigons in L that contain a_1, a_2, a_6 and a_5 respectively. Let $\mathcal{Q}$ be the partition $\{\{B_1\}, \{B_2, B_3\}, \{B_4\}\}$ of $\mathcal{B}(L)$. It is clear that L is partially congruent to K with respect to the partition $\mathcal{Q}$. Furthermore L is faithful to K with respect to $\mathcal{S}$. We note that B_2 and B_3 belong to distinct components of L, in agreement with Lemma 6.

Let $\mathcal{S}$ be a semicycle cover in a 3-graph K. The semicycles of $\mathcal{S}$ induce a partition of VK into pairs, where two vertices belong to the same cell in this partition if and only if they are the two poles of a semicycle in $\mathcal{S}$ with respect to some red-yellow bigon. Let $\mathcal{V}$ denote this partition. Let L be the 3-graph obtained from K by deleting the yellow edges and inserting a yellow edge joining the vertices v and w for each $\{v, w\} \in \mathcal{V}$. We say that L *encodes* K (with respect to $\mathcal{S}$). Clearly L is partially congruent to K with respect to a partition $\mathcal{Q}$. We say that $\mathcal{Q}$ is the *encoding partition.* Furthermore we have a one to one correspondence between the semicycles in $\mathcal{S}$ and the blue-yellow bigons in L. Hence L is faithful to K with respect to $\mathcal{S}$.

EXAMPLE 7. Returning to Example 6, we see that L is in fact the 3-graph that encodes K with respect to $\mathcal{S}$, and that $\mathcal{Q}$ is the encoding partition. We also note that L is spherical. Lemma 7 below states that L will be spherical in general whenever $\mathcal{S}$ is a spanning semicycle cover.

LEMMA 7. *Let L be the 3-graph that encodes a 3-graph K with respect to some spanning semicycle cover $\mathcal{S}$. Then L is spherical.*

PROOF. Let C be a circuit in L. By Lemma 5, $\beta(C)$ is the blue edge set of a cycle in K. Since $\mathcal{S}$ is spanning, then $\beta(C)$ is the blue edge set of a cycle $\sum(\mathcal{U} \cup \mathcal{Y})$ for a set $\mathcal{U}$ of semicycles in $\mathcal{S}$ and a set $\mathcal{Y}$ of red-blue bigons in K. Let $\mathcal{R}$ denote the set of blue-yellow bigons in L that correspond to the semicycles in $\mathcal{U}$. Evidently $\beta(C)J = J\beta(\sum(\mathcal{U} \cup \mathcal{Y})) = \beta(\sum(\mathcal{R} \cup \mathcal{Y}))$, and $\sum(\mathcal{R}J \cup \mathcal{Y})$ is a cycle in L. By Lemma 1, $C = \sum(\mathcal{B} \cup \mathcal{R} \cup \mathcal{Y})$ for some set $\mathcal{B}$ of red-yellow bigons in L. Hence C is a sum of bigons, and we conclude that L is spherical. $\square$

Let A and B be distinct red-yellow bigons of a 3-graph K. Let c_1 be a yellow edge in A and c_2 a yellow edge in B. Let c_1 join v_1 and w_1 and let c_2 join v_2 and w_2. Let L be the 3-graph obtained from K by deleting c_1 and c_2 and inserting two new yellow edges that join v_1 to v_2 and w_1 to w_2 respectively. We say that L is a 3-graph obtained by *coalescing* A and B. If A and B belong to distinct components of K, then clearly $c(L) = c(K) - 1$.

The following lemma is immediate.

LEMMA 8. *Suppose L is partially congruent to a 3-graph K with respect to a partition $\mathcal{Q}$. Furthermore suppose J is obtained from L by coalescing two red-yellow bigons that belong to the same cell in $\mathcal{Q}$. Then J is partially congruent to K.*

LEMMA 9. *Let A and B be red-yellow bigons that belong to distinct components of a 3-graph K. Let L denote a 3-graph obtained from K by coalescing A and B. Then $\chi(L) = \chi(K) - 2$. Hence L is spherical if and only if K is spherical.*

PROOF. Clearly $|VL| = |VK|$. However the number of red-yellow bigons has dropped by 1 as has the number of blue-yellow bigons. Hence

$$\begin{aligned} \chi(L) &= |\mathcal{B}(K)| + |\mathcal{R}(K)| + |\mathcal{Y}(K)| - 2 - \frac{|VK|}{2} \\ &= \chi(K) - 2. \quad \square \end{aligned}$$

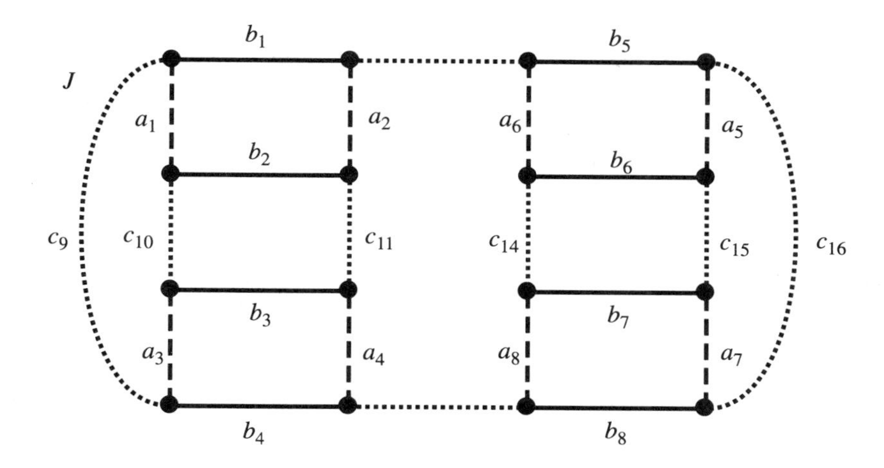

FIGURE 11.

EXAMPLE 8. Consider the 3-graphs K and L of Example 6. The 3-graph J in Figure 11 is a 3-graph obtained from L by coalescing B_2 and B_3. We note that J is a spherical 3-graph congruent to K.

THEOREM 4. *If there exists a spanning semicycle cover $\mathcal{S}$ in a 3-graph K, then K is congruent to a spherical 3-graph.*

PROOF. Let L_1 be the 3-graph that encodes K with respect to $\mathcal{S}$. Let $\mathcal{Q}_1$ be the encoding partition. By Lemma 7 L_1 is spherical. Recall that L_1 is partially congruent to K with respect to $\mathcal{Q}_1$ and L_1 is faithful to K. If each cell of $\mathcal{Q}_1$ is a singleton, then L_1 is congruent to K and we are done.

Now suppose that there are two distinct red-yellow bigons B_1 and B_2 that belong to a common cell of $\mathcal{Q}_1$. Since $\mathcal{S}$ is a spanning semicycle cover, then by Lemma 6 B_1 and B_2 belong to distinct components of L_1. Let L_2 denote a 3-graph obtained from L_1 by coalescing B_1 and B_2. By Lemma 9, L_2 is spherical. By Lemma 8, L_2 is partially congruent to K with respect to a partition $\mathcal{Q}_2$. Furthermore, for any $C \in \mathcal{S}$, $|C \cap \partial V B_1|$ and $|C \cap \partial V B_2|$ are both even, and therefore $|C \cap \partial V B|$ is even for all $B \in \mathcal{B}(L_2)$. Hence we conclude that L_2 is faithful to K.

Proceeding inductively, we obtain a spherical 3-graph L_n partially congruent to K with respect to a partition $\mathcal{Q}_n$ such that each cell in $\mathcal{Q}_n$ is a singleton. Hence L_n is congruent to K. $\square$

6. MacLane's Theorem

A family $(C_1, C_2, \ldots, C_n)$ of cycles in a graph G is a *cycle double cover* of G if for each edge $e \in EG$, e belongs to C_i for exactly two values of i, where $1 \leq i \leq n$.

A family of cycles in a graph G is said to be a *spanning family* if its components span $\mathcal{Z}(G)$. Similarly we may talk about a *spanning set* of cycles of G. If the cycles of the spanning family or spanning set constitute a cycle double cover, then this double cover is also described as *spanning*. Similarly we also talk about a *spanning circuit double cover*.

The *foundation* of a graph G is the subgraph spanned by the complement in EG of the set of isthmuses of G. In this section we specialise Theorem 1 to the case of gems to obtain MacLane's theorem. If C is a cycle in a 3-graph K, then we denote by $N_Y(C)$ the set of all red-blue bigons that meet $\beta(C)$.

THEOREM 5. *[MacLane]. A graph is planar if and only if its foundation has a spanning circuit double cover.*

PROOF. Firstly, let G be a planar graph. Therefore G underlies a spherical gem K. By Theorem 1 there exists a spanning semicycle cover $\mathcal{S}$ in K. Suppose there is a semicycle $S \in \mathcal{S}$ that contains both blue edges of a red-blue bisquare Y. Therefore S corresponds to a path $\{e\}$ of length 1 in G. Assume that e is not an isthmus and hence that there exists a circuit C in G contains e. Let D denote a semicycle in K that represents C. Then $\mathcal{U}$ contains only one blue edge of Y. Since no circuit in $\mathcal{B}(K) \cup \mathcal{Y}(K) \cup \mathcal{S}$ contains just one blue edge of Y we have a contradiction to the fact that $\mathcal{B}(K) \cup \mathcal{Y}(K) \cup \mathcal{S}$ spans $\mathcal{Z}(K)$. Hence we conclude that each semicycle of $\mathcal{S}$ represents a circuit or corresponds to an isthmus in G. Let $S' = \{S_1, S_2, \ldots, S_n\}$ denote the set of semicycles in $\mathcal{S}$ that represent circuits in G. Clearly a semicycle in $\mathcal{S} - \mathcal{S}'$ is the sum of a red-blue bisquare and red-yellow bigons. Hence $\mathcal{B}(K) \cup \mathcal{Y}(K) \cup \mathcal{S}'$ spans the cycle space of K.

Let C_i be the circuit of G that corresponds to S_i for each i. Since the set of intersections of members of $\mathcal{S}$ with $\beta(K)$ is a partition of $\beta(K)$, each red-blue bigon that does not correspond to an isthmus belongs to $N_Y(S_i)$ for exactly two values of i. It then follows that each edge in the foundation of G belongs to C_i for exactly two values of i. Hence $\mathcal{C} = (C_1, C_2, \ldots, C_n)$ is a circuit double cover for the foundation of G. We are required to show that $C_1, C_2, \ldots, C_n$ span $\mathcal{Z}(G)$. Let D be a circuit in G. Let S be a semicycle in K that represents D. Hence $S = \sum \mathcal{U}$ for a set U of circuits in $\mathcal{B}(K) \cup \mathcal{Y}(K) \cup \mathcal{S}'$. Evidently no red-blue bigon has both of its blue edges in S, for otherwise S would not represent a circuit in G. Therefore $S + \sum \mathcal{U}_1$ is also a semicycle that represents D, where $\mathcal{U}_1$ is any set of red-yellow bigons in $N(S)$ and red-blue bigons in $N_Y(S)$. Let $\mathcal{U}_2$ be the set of red-blue bigons in $\mathcal{U}$ that are not in $N_Y(S)$. Then by choosing $\mathcal{U}_1$ appropriately we find that $\sum \mathcal{U}_2 + \sum(\mathcal{S}' \cap \mathcal{U}) + \cup \mathcal{B}$ is also a semicycle that represents D, where $\mathcal{B}$ is the set of red-yellow bigons included in $\sum \mathcal{U}_2 + \sum(\mathcal{S}' \cap \mathcal{U})$. Let V be the set of components of $\mathcal{C}$ that correspond to the semicycles in $\mathcal{S}' \cap \mathcal{U}$. Then it follows that $\sum \mathcal{V} = D$ since the members of $\mathcal{U}_2$ are not in $N_Y(S)$. We conclude that the compo C span $\mathcal{Z}(G)$, as required.

Now suppose the foundation of G has a spanning circuit double cover

$$(C_1, JC_2, \ldots, C_m).$$

Let K be a gem that G underlies. Let S_i be a semicycle in K that represents C_i, and let

$$\mathcal{S} = \{S_1, S_2, \ldots, S_m\} \cup \{S_{m+1}, S_{m+2}, \ldots, S_n\},$$

where $\{S_{m+1}, S_{m+2}, \ldots, S_n\}$ is the set of red-blue bigons in K that correspond to the isthmuses in G. Since $S_i + Y$ is a semicycle that also represents C_i, where $i \leq m$ and $Y \in N_Y(S_i)$, and each red-blue bigon not in $\{S_{m+1}, S_{m+2}, \ldots, S_n\}$ belongs to $N_Y(S_i)$ for exactly two values of $i \leq m$, we may choose each S_i so that the set of intersections of members of S with $\beta(K)$ is a partition of $\beta(K)$.

We now show that $\mathcal{B}(K) \cup \mathcal{Y}(K) \cup \mathcal{S}$ spans $\mathcal{Z}(K)$. Let D be a cycle in K. Suppose Y is a red-blue bigon such that both blue edges of Y belong to D. Then $D + Y$ is a sum of circuits in $\mathcal{B}(K) \cup \mathcal{Y}(K) \cup \mathcal{S}$ if and only if D is a sum of circuits in $\mathcal{B}(K) \cup \mathcal{Y}(K) \cup \mathcal{S}$. We therefore assume that D does not contain both blue edges of a red-blue bigon. Since D is a cycle, the number of blue edges of D incident on a given red-yellow bigon is always even. Therefore the set of edges that correspond to the bigons of $N_Y(D)$ is a cycle C Since $C_1, C_2, \ldots, C_m$ span $\mathcal{Z}(G)$ then $C = \sum \mathcal{V}$ for some set $\mathcal{V} \subseteq \{C_1, C_2, \ldots, C_m\}$. Let $\mathcal{U}$ be the set of semicycles in $\mathcal{S}$ that represent the circuits of $\mathcal{V}$. Let $D_1 = \sum \mathcal{U}$. Let $D_2 = D_1 + \sum \mathcal{U}_3$ where $\mathcal{U}_3$ is the set of red-blue bigons that have both blue edges in D_1. Evidently D_2 is a cycle such that $N_Y(D_2) = N_Y(D)$ and is a sum of circuits in $\mathcal{B}(K) \cup \mathcal{Y}(K) \cup \mathcal{S}$. By adding to D_2 bigons in $N_Y(D_2)$ we can obtain a cycle D_3 such that $\beta(D_3) = \beta(D)$. By Lemma 1, the fact that D_3 is a sum of circuits in $\mathcal{B}(K) \cup \mathcal{Y}(K) \cup \mathcal{S}$ implies that D is also a sum of circuits in $\mathcal{B}(K) \cup \mathcal{Y}(K) \cup \mathcal{S}$. We conclude that $\mathcal{S}$ is a spanning semicycle cover in K. By Theorem 1 K is congruent to a spherical gem K'. Evidently G underlies K' and hence G is planar. □

References

1. Craig Paul Bonnington and Charles H. C. Little, *The classification of combinatorial surfaces using 3-graphs*, Australasian Journal of Combinatorics **5** (1992).
2. M. Ferri and C. Gagliardi, *Crystallization moves*, Pacific J. Math **100** (1982), 85–103.
3. M. Ferri, C. Gagliardi, and L. Grasselli, *A graph-theoretical representation of pl-manifolds—a survey on crystallizations*, Aequationes Mathematicae **31** (1986), 121–141.
4. S. Lins, *Graphs of maps*, Ph.D. thesis, University of Waterloo, 1980.
5. ———, *Graph-encoded maps*, J. Combin. Theory Ser. B **32** (1982), 171–181.
6. Charles H. C. Little, *Cubic combinatorial maps*, J. Combin. Theory Ser. B **44** (1988), 44–63.
7. Charles H. C. Little and Andrew Vince, *Embedding schemes and the jordan curve theorem*, Topics in Combinatorics and Graph Theory (Heidelberg) (R. Bodendiek and R. Henn, eds.), Physica- Verlag, Heidelberg, 1990, pp. 479–489.
8. Saunders MacLane, *A combinatorial condition for planar graphs*, Fund. Math. **28** (1937), 22–32.
9. N. Robertson, *Graphs minimal under girth, valency and connectivity*, Ph.D. thesis, University of Waterloo, 1971.

10. Saul Stahl, *A combinatorial analog of the Jordan curve theorem*, J. Combin. Theory Ser. B **35** (1983), 28–38.
11. Andrew Vince, *The classification of closed surfaces using coloured graphs*, to appear, 1990.

(C. P. Bonnington) Department of Mathematics and Statistics, University of Auckland, Private Bag, Auckland, New Zealand
E-mail address: bonning@mat.aukuni.ac.nz

(C. H. C. Little) Department of Mathematics, Massey University, Palmerston North, New Zealand
E-mail address: C.Little@massey.ac.nz

DIMACS Series in Discrete Mathematics
and Theoretical Computer Science
Volume 9, 1993

Chordal completions of grids and planar graphs

F. R. K. CHUNG AND DAVID MUMFORD

September 17, 1992

ABSTRACT. It is common to model a finite probability space with a graph where nodes correspond to events and edges indicate dependent pairs of events. This paper is an extended abstract of a full length article in which we study chordal completions of graphs which are related to models in which marginal and conditional probabilities can be efficiently computed.

1. Introduction

Increasingly, artificial intelligence has turned to the theory of Bayesian statistics to provide a solid theoretical foundation and a source of useful algorithms for reasoning about the world in conditions of uncertain and incomplete information. This is true both in familiar high-level applications, such as medical expert systems, and in low-level applications such as speech recognition and computer vision (see **[8]** for an overview, **[6]** for medical applications, **[9]** for speech, **[5]** for vision). However, all serious applications demand probability spaces with thousands of random variables, and some simplification is required before you can even write down probability distributions in such spaces. The reason this approach is even partially tractable is that one assumes there are many pairs of random variables which are conditionally independent, given various other variables. One extremely useful way to describe this sort of probability space is based on graph theory: one assumes that a graph G is given, whose vertices $V(G)$ correspond to the random variables in the application, and whose edges $E(G)$ denote pairs of variables which *directly* affect each other. What this means is that if $v, w \in V(G), S \subset V(G)$, and every path from v to w crosses S, then the corresponding variables X_v, X_w are conditionally independent given X_S. As is well-known, this assumption implies that the probability distribution has the

1991 *Mathematics Subject Classification.* 05C35.
This is an extended abstract of a paper which will be submitted for publication elsewhere.

Gibbs form:

$$\Pr(\vec{X}_v) = \frac{e^{-\sum_C E_C(\{X_w\}_{w \in C})}}{Z}$$

where C runs over the *cliques* of G containing the vertex v, E_C is a measure of the likelihood of the simultaneous values of the variables in the clique C, and Z is a normalizing constant.

A typical problem in this setting is to find the maximum likelihood estimate of the variables $\vec{X}_v$, i.e. the minimum of the so-called energy

$$E(\vec{X}_v) = \sum_C E_C(\{X_w\}_{w \in C}).$$

Unfortunately, minimizing such complex functions of huge numbers of variables is not an easy task. One situation in which the minimum can be quickly and accurately computed is that studied in dynamic programming [**1**]. This is the case where the variables can be ordered in such a way that X_k is conditionally independent of all but a few of the previous X_l's, given the values of these few. A Markov chain is the simplest example of this, and this approach, under the name of the Viterbi algorithm, dominates research in speech recognition. However, it has turned out that modifications of the dynamic programming perspective are much more widely applicable [**6**]. In [**6**], the authors propose using a *chordal completion* $\hat{G}$: this is a chordal graph with the same vertex set $V(G)$ and edge set containing $E(G)$. Recall that a chordal graph is a graph in which all cycles of length at least 4 contain chords, sometimes called a triangulated graph. (For graph-theoretical terminology, the reader is referred to [**2**].) If the cliques in $\hat{G}$ are not too large, one can carry out a variant of dynamic programming for Gibbs fields based on G, and compute essentially all marginal and conditional probabilities of interest.

2. Grid Graphs

In computer vision, one seeks to analyze a two-dimensional signal, finding first edges and areas of homogeneous texture, secondly using these to segment the domain of the signal and thirdly identifying particular regions as resulting from the play of light and shadow on known types of objects such as faces. The random variables that arise in this analysis are firstly I_{ij}, the light intensity measured by a receptor at a position (i, j) of the camera's or eye's focal plane, secondly "line processes" l_{ij} indicating an edge separating adjacent "pixels" (i, j) and $(i, j+1)$ or $(i+1, j)$, and many higher level variables. What interests us is that the measured variables are parametrized by points of a lattice, and that the structures which one calculates are found by examining *local* interactions of these variables. In fact, even a high level variable like the presence of a face is linked to local areas of the image, rather than the whole image, because a face will usually be a subset of the image domain and its presence is more or less independent of the scene in the background. What this means is that the cliques of the graph involve local areas in the lattice, and do not require long range interaction of the pixel values I_{ij}. The simplest example of such a graph is the simple n by n grid, which we denote by L_n: it has vertex set

$$V(L_n) = \{(i,j) : 0 \leq i,j \leq n\}$$

and edges joining:

$$(i,j) \text{ to } (i+1,j), 0 \leq i < n, 0 \leq j \leq n,$$
$$(i,j) \text{ to } (i,j+1), 0 \leq i \leq n, 0 \leq j < n.$$

What we would like to know is how big are the chordal completions of graphs of this sort: how many edges do they have and what are their degrees? We prove the following theorems for grid graphs:

THEOREM 1. *A chordal completion of the n by n grid L_n must contain at least $c\,n^2 \log n$ edges for some constant c. Furthermore, we construct a chordal completion of L_n with $(7.75)n^2 \log n$ edges.*

We note that throughout this paper, all logarithms are to the base 2. By using results on the treewidth of a graph[**10**], we prove

THEOREM 2. *A chordal completion of the n by n grid must contain a vertex of degree cn for some absolute constant c.*

The above theorems can be generalized to all planar graphs by using the planar separator theorems [**4, 7**].

THEOREM 3. *A planar graph on n vertices has a chordal completion with $cn \log n$ edges for some absolute constant c.*

It is also known that there is an $O(n\ log\ n)$ algorithm for constructing the chordal completion of a planar graph.

How good are these bounds? Compared to random graphs, they look quite good: P. Erdös first raised the question how large is a chordal completion of a random graph on n vertices with edge density k/n for some fixed k (or random k-regular graphs). It turns out that chordal completions of random graphs must contain cn^2 edges for some constant c. The reader is referred to [**2**] for models of random graphs or random regular graphs. Unfortunately for the application to computer vision, the lower bounds on the size of the chordal completions of grids are still too big to make the use of dynamic programming or its variants practical in vision: typical values of n are 100 or more, and probability tables for the values of 100 random variables are quite impossible. However, the construction given in the full paper for a chordal completion of L_n is strongly reminiscent of the approach to vision problems called "pyramid algorithms" [**11**], e.g. wavelet expansions [**3**]. This link is interesting to explore.

References

1. R. Bellman, *Dynamic Programming*, Princeton Univ. Press (1957).
2. B. Bollobás, Random Graphs, *Academic Press,* New York (1987)
3. P. Burt and E. Adelson, The Laplacian pyramid as a compact image code, *IEEE Trans. on Comm.*, **31** (1983), 532-540.
4. F. R. K. Chung, Improved separators for planar graphs, in *Graph Theory, Combinatorics, and Applications*, Y. Alavi et al eds., (1991), 265-270.
5. R. Geman, Random Fields and Inverse Problems in Imaging, Springer Lecture Notes in Mathematics.
6. Lauritzen and Spiegelhalter, Local computations with probabilities on graphical structures and their applications to expert systems, *J.Royal Soc., Part B*, **50** (1988) 157-224.
7. R. J. Lipton and R. E. Tarjan, A separator theorem for planar graphs, *SIAM J. Appl. Math.* **36** (1979) 177-189
8. J. Pearl, *Probabilistic reasoning in intelligent systems: networks of plausible inference*, Morgan Kaufmann (1988).
9. L. Rabiner, A Tutorial on Hidden Markov Models and Selected Applications in Speech Recognition, *Proc. of the IEEE*, **77** (1989), 257-285.
10. N. Robertson and P. D. Seymour, Graph minors II, Algorithmic aspects of tree-width, *J. Algorithms* **7**(1986), 309-322
11. A. Rosenfeld ed., *Multiresolution Image Processing and Analysis*, Springer Verlag, 1984.
12. J. R. Walter, Representation of rigid cycle graphs, Ph. D. Thesis, Wayne State University 1972

BELLCORE, MORRISTOWN, NJ 07962
E-mail address: frkc@bellcore.com

HARVARD UNIVERSITY, CAMBRIDGE, MA 02138
E-mail address: mumford@sancho.harvard.edu

DIMACS Series in Discrete Mathematics
and Theoretical Computer Science
Volume 9, 1993

Upward Planar Drawing of Single Source Acyclic Digraphs

MICHAEL D. HUTTON AND ANNA LUBIW

September 30, 1992

ABSTRACT. An upward plane drawing of a directed acyclic graph is a plane drawing of the graph in which each directed edge is represented as a curve monotone increasing in the vertical direction. Thomassen [20] has given a non-algorithmic, graph-theoretic characterization of those directed graphs with a single source that admit an upward plane drawing. We present an efficient algorithm to test whether a given single-source acyclic digraph has an upward plane drawing and, if so, to find a representation of one such drawing.

The algorithm decomposes the graph into biconnected and triconnected components, and defines conditions for merging the components into an upward plane drawing of the original graph. To handle the triconnected components we provide a linear algorithm to test whether a given plane drawing of a single source digraph admits an upward plane drawing with the same faces and outer face, which also gives a simpler, algorithmic proof of Thomassen's result. The entire testing algorithm (for general single-source directed acyclic graphs) operates in $O(n^2)$ time and $O(n)$ space, and represents the first polynomial time solution to the problem.

1. Introduction

There are a wide range of results dealing with drawing, representing, or testing planarity of graphs. Fáry [9] showed that every planar graph can be drawn in the plane using only straight line segments for the edges. Tutte [21] showed that every 3-connected planar graph admits a convex straight-line drawing, where the facial cycles other than the unbounded face are all convex polygons. The first

1991 *Mathematics Subject Classification.* 05C85, 68Q20, 05C10, 06A07.

Supported in part by NSERC. This paper is an extended abstract of a full length journal article submitted for publication elsewhere. A preliminary discussion of the result appeared in SODA 91. This update gives a more general combinatorial characterization.

linear time algorithm for testing planarity of a graph was given by Hopcroft and Tarjan [**11**].

Planar graph layout has many interesting applications in automated circuit design, for representation of network flow problems (e.g. PERT graphs in software engineering) and artificial intelligence.

An *upward plane drawing* of a digraph is a plane drawing such that each directed arc is represented as a curve monotone increasing in the y-direction. In particular the graph must be a directed acyclic graph (DAG). A digraph is *upward planar* if it has an upward plane drawing. Consider the digraphs in Figure 1. By convention, the edges in the diagrams in this paper are directed upward unless specifically stated otherwise, and direction arrows are omitted unless necessary. The digraph on the left is upward planar: an upward plane drawing is given. The digraph on the right is not upward planar—though it is planar, since placing v inside the face f would eliminate crossings, at the cost of producing a downward edge.

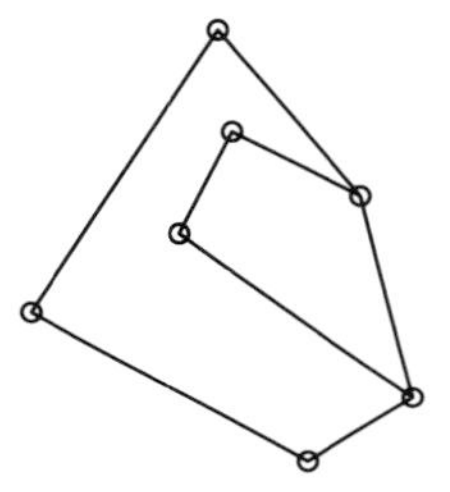

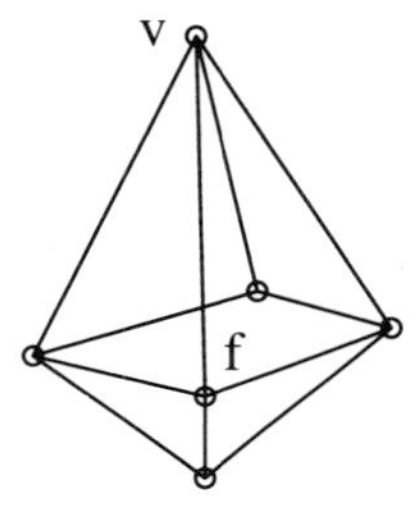

FIGURE 1. Upward planar and non-upward planar graphs.

Kelly [**14**] and Kelly and Rival [**15**] have shown that for every upward plane drawing there exists a *straight-line* upward plane drawing with the same faces and outer face, in which every edge is represented as a straight line segment. This is an analogue of Fáry's result for general planar graphs. The general problem of recognizing upward planar digraphs is not known to be in P, nor known to be NP-hard. For the case of single-source single-sink digraphs there is a polynomial time recognition algorithm provided by Platt's result [**16**] that such a graph is upward planar iff the graph with a source-to-sink edge added is planar. An algorithm to find an upward plane drawing of such a graph was given by DiBattista and Tamassia [**6**]. For the special case of bipartite graphs, upward planarity is equivalent to planarity [**5**].

In this paper we will give an efficient algorithm to test upward planarity for single-source digraphs, eliminating the single-sink restriction. For the most part we will be concerned only with constructing an upward planar *representation*—enough combinatorial information to specify an upward plane drawing without giving actual numerical coordinates for the vertices. This notion will be made precise in Section 3. We will remark on the extension to a drawing algorithm in Section 7. Our main result is an $O(n^2)$ algorithm to test whether a given single-source digraph is upward planar, and if so, to give a representation for it which leads to a drawing with known methods. This result is partly based on a graph-theoretic result of Thomassen [**20**, Theorem 5.1]:

THEOREM 1.1 (THOMASSEN). *Let Γ be a plane drawing of a single-source digraph G. Then there exists an upward plane drawing Γ' strongly equivalent to (i.e. having the same faces and outer face as) Γ if and only if the source α of G is on the outer face of Γ, and for every cycle Σ in Γ, Σ has a vertex β which is not the tail of any directed edge inside or on Σ.*

The necessity of Thomassen's condition is clear: for a graph G with upward plane drawing Γ', and for any cycle Σ of Γ', the vertex of Σ with highest y-coordinate cannot be the tail of an edge of Σ, nor the tail of an edge whose head is inside Σ.

Since a 3-connected graph has a unique planar embedding (up to the choice of the outer face) by Whitney's theorem (cf. [**2**]) Thomassen concludes that his theorem provides a "good characterization" of 3-connected upward planar graphs— i.e. puts the class of 3-connected upward planar graphs in NP intersect co-NP. An efficient algorithm is not given however (there are potentially an exponential number of possible cycles to check), nor does Thomassen address the issue of non-3-connected graphs (which could have an exponential number of different planar embeddings).

The problem thus decomposes into two main issues. The first is to describe Thomassen's result algorithmically; we do this in Section 4 with a linear time algorithm, which provides an alternative proof of his theorem. The second issue is to isolate the triconnected components of the input graph, and determine how to put the "pieces" back together after the embedding of each is complete. This more complex issue is treated in Section 6, after a discussion of decomposition properties in Section 5.

The algorithm for splitting the input into triconnected components and merging the embeddings of each operates in $O(n^2)$ time. Since a triconnected graph is uniquely embeddable in the plane up to the choice of the outer face, and the number of possible external faces of a planar graph is linear by Euler's formula, the overall time to test a given triconnected component is also $O(n^2)$, so the entire algorithm is quadratic.

2. Preliminaries

In addition to the definitions below we will use standard terminology and notation of Bondy and Murty [**2**].

All graphs in this paper are directed and acyclic unless otherwise stated. We will use the term *cycle* and the various notions of connectivity with respect to the *underlying undirected graph*, so a digraph G is *connected* if there exists an undirected path between any two vertices in G. For S a set of vertices,

$G \backslash S$ denotes G with the vertices in S and all edges incident to vertices in S removed. If S contains a single vertex v we will use the notation $G \backslash v$ rather than $G \backslash \{v\}$. G is k-connected if it has at least $k+1$ vertices and the removal of at least k vertices is required to *disconnect* the graph. By Menger's Theorem [**2**], G is k-connected if and only if there exist k vertex-disjoint undirected paths between any two vertices. A set of vertices whose removal disconnects the graph is a *cut-set*. The terms *cut vertex* and *separation pair* apply to cut-sets of size one and two respectively. A graph which has no cut vertex is *biconnected* (2-connected). A graph with no separation pair is *triconnected* (3-connected). For G with cut vertex v, a *component* of G with respect to v is formed from a connected component H of $G \backslash v$ by adding to H the vertex v and all edges between v and H. For G with separation pair $\{u, v\}$, a *component* of G with respect to $\{u, v\}$ is formed from a connected component H of $G \backslash \{u, v\}$ by adding to H the vertices u, v and all edges between u, v and vertices of H. The edge (u, v), if it exists, forms a component by itself. An algorithm for finding triconnected components[1] in linear time is given in Hopcroft and Tarjan [**12**]. A related concept is that of *graph union*: we define $G_1 \cup G_2$, for components with "shared" vertices to be the *inclusive* union of all vertices and edges. That is, for v in both G_1 and G_2, the vertex v in $G_1 \cup G_2$ is adjacent to edges in each of the subgraphs G_1 and G_2.

Contracting an edge $e = (u, v)$ in G results in a graph, denoted G/e, with the edge e removed, and vertices u and v *identified*. Inserting new vertices within edges of G generates a *subdivision* of G. A *directed subdivision* of a digraph results from repeatedly adding a new vertex w to divide an edge (u, v) into (u, w) and (w, v). G_1 and G_2 are *homeomorphic* if both are subdivisions of some other graph. G is planar if and only if every subdivision of G is planar [**2**].

In a directed graph, the *in-degree* of a vertex v is the number of edges directed towards v, denoted $deg^{-}v$. Analogously the *out-degree* ($deg^{+}v$) of v is the number of edges directed away from v. A vertex of in-degree 0 is a *source* in G, and a vertex of out-degree 0 is a *sink*.

Adopting some poset notation: we will write $u \leq v$ is there is a directed path $u \xrightarrow{*} v$ of length 0 or more, and $u < v$ ($u \xrightarrow{+} v$) to emphasize that u and v are distinct. Vertices u and v are *comparable* if $u \leq v$ or $v \leq u$, and *incomparable* otherwise. If (u, v) is an edge of a digraph then u *dominates* v, u is *incident to* v, and v is *incident from* u.

3. A Combinatorial View of Upward Planarity

As discussed by Edmonds and others (see [**10**]) a connected graph G is planar iff it has a *planar representation*: a cyclic ordering of edges around each vertex such that the resulting set of *faces* F satisfies $2 = |F| - |E| + |V|$ (Euler's formula). A *face* is a cyclically ordered sequence of edges and vertices $v_0, e_0, v_1, e_1, \ldots, v_{k-1}, e_{k-1}$, where $k \geq 3$, such that for any $i = 0, \ldots, k-1$

[1] Note that Hopcroft and Tarjan's "components" include an extra (u, v) edge.

the edges e_{i-1} (subscript addition modulo k) and e_i are incident with the vertex v_i and consecutive in the cyclic edge ordering for v_i.

We will say that two plane drawings are *equivalent* if they have the same representation—i.e. the same set of faces. Two plane drawings are *strongly equivalent* if they have the same representation and the same outer face.

One method of combinatorially specifying an upward planar drawing is provided by the following result of (independently) DiBattista and Tamassia [**6**], and Kelly [**14**]. They use the concept of a planar *s-t graph*, defined to be a planar DAG which has a single source s, a single sink t and contains the edge (s, t)—exactly the upward planarity condition of Platt [**16**] for single-source single-sink digraphs.

THEOREM 3.1 (DIBATTISTA AND TAMASSIA, KELLY). *Let G be a directed acyclic graph. If G is upward planar, then edges can be added to it to obtain a planar s-t graph (i.e. G is a (spanning) subgraph of a planar s-t graph). Conversely, if edges can be added to G to obtain a planar s-t graph G', then G is upward planar. Furthermore, for any planar embedding Γ of G' with (s, t) on the outer face, there is an upward plane drawing of G strongly equivalent to Γ with the extra edges removed.*

The final statement was not explicitly given, however to prove their result, DiBattista and Tamassia give an algorithm which takes a planar s-t graph, finds an *arbitrary* planar representation of it and outputs an upward plane drawing which respects this embedding, so the statement follows. Their algorithm, which we will require later in the paper, runs in $O(n \log n)$ arithmetic steps.

The disadvantage of this NP characterization in terms of planar s-t graphs is the difficulty of testing it. Thomassen's co-NP condition on single-source digraphs suffers from the same problem. For the case of single-source digraphs, we will give a testable (algorithmic) characterization in the next section.

To provide some motivation for this algorithm, we give another characterization of single-source upward planar graphs, equivalent to Thomassen's:

First we define $P(v)$, the *predecessor set of* v to be the set $\{u : u \leq v \text{ in } G\}$. Notice the set $P(v)$ includes v. Define G_v to be induced subgraph of G on $P(v)$. For a planar representation Γ of G, define Γ_v to be the planar representation induced by Γ on G_v.

PROPOSITION 3.1. *Given a single source DAG G, and a planar representation Γ of G with a specified outer face and source s on the outer face, G has an upward plane drawing strongly equivalent to Γ* iff *the following condition holds:*

CONDITION 3.1. *For each vertex $v \in V$, v is a sink on the outer face of the planar embedding Γ_v induced by $P(v)$.*

We will often refer to a planar representation Γ satisfying Condition 3.1 as an *upward planar representation* of G.

Since a strongly equivalent upward plane drawing provides the same planar representation Γ and predecessors of v have smaller y-coordinates in the drawing, v must be on the outer face of Γ'_v. Thus the necessity holds. For sufficiency, it can be shown that the existence of a violating cycle (defined below) arising from Thomassen's characterization implies the existence of some vertex inside the cycle for which Condition 3.1 cannot hold.

Define a *violating cycle* of G with respect to Γ to be a cycle Σ such that every vertex of Σ is the tail of an edge inside or on Σ. This is exactly the cycle Σ from Theorem (1.1). As observed in the introduction, a violating cycle in Γ precludes the existence of an strongly equivalent upward drawing.

The results of this section, combined with the characterization of DiBattista and Tamassia, and the single-source characterization of Thomassen give:

THEOREM 3.2. *The following conditions are equivalent for a single-source DAG G with planar representation Γ having a designated outer face and single source s which is on the outer face:*

(i) *G has an upward plane drawing strongly equivalent to Γ.*

(ii) *G is a (spanning) subgraph of some planar s-t graph which has an upward plane drawing strongly equivalent to Γ (after removal of the extra edges).*

(iii) *for all $v \in G$, v is a sink on the outer face of Γ_v.*

(iv) *Γ does not contain a violating cycle.*

We note that condition (iii) is the only one which can obviously be tested in polynomial time, and provide a linear time algorithm in the next section.

4. Strongly-Equivalent Upward Planarity

Consider the following question: Given a single-source acyclic digraph G and a planar representation Γ for G, with s on the outer face of Γ, does G admit an upward planar drawing strongly equivalent to Γ?

Here, we present a linear time algorithm to test whether G has an upward planar embedding strongly equivalent to Γ with a designated outer face. The algorithm will return the edges necessary to augment G so that sinks occur only on the outer face in the positive case, or a violating cycle in the negative case. Since any planar representation of a single-source DAG with the source and all sinks on the outer face is a subgraph of a planar s-t graph—simply designate one sink as t, and add an edge from the source and all other sinks to it—the algorithm provides a new proof of Thomassen's theorem.

The algorithm is recursive, and the proof that it works is by induction. If there is a sink v on the outer face of Γ, then recursively (trivial if G has one node) determine a violating cycle for $G\backslash v$ (in which case we are done) or a set of edges X required to augment $\Gamma\backslash v$ ($G\backslash v$) to a planar representation Γ' with all sinks on the outer face. Now add v and edges incident to v to the outer face of Γ'. To determine the additional required edges to resolve the internal sinks in the new faces, consider all vertices w which are sinks on the outer face of Γ', but are not sinks on the outer face of Γ. Adding the edges (w, v) (where they do not already exist in G) to X retains planarity, single-sourcedness and acyclicity in $G \cup X$ and does not change the outer face.

It remains to deal with the case when the outer face of Γ has no sink. We claim that in this case G has a violating cycle: If the outer face of Γ is a cycle then it is a violating cycle. If the outer face is a walk, then follow it starting at s, and let v be the first vertex which repeats. Vertex v must be a cut vertex. Consider the segment of the walk from v to v. If this segment contains only one other vertex, say u, then u is a sink, contradiction. Otherwise we obtain a cycle C from v to v. The two edges incident with v must be directed away from v and no other vertex is a sink on C, so C must be a violating cycle. □

The above algorithm can be implemented in linear time (so that each vertex is involved in no more than a constant number of operations), using data structures no more complicated than a linked list.

5. Decomposition properties of Upward Planar Graphs

Here we describe some properties which preserve upward planarity. The purpose is twofold: firstly, the properties are necessary for the proofs in the next section; secondly, they provide an intuitive look at the structure of upward planar graphs, and hence motivate the decomposition approach we take in the recognition algorithm. Though the proofs are not shown the reader should be somewhat convinced after viewing Figure 2.

LEMMA 5.1. *We have the following properties of upward planar graphs:*

(*) *Any connected upward planar G is a subgraph of some single-source upward planar G^* such that all non-source v in G have the same in-degree in G^* as in G.*

(a) *Let G be a DAG, and v dominated by u be a vertex of G with in-degree 1. Then, $G/(u, v)$ is upward planar if G is upward planar. (See Figure 2(a).)*

(b) *Let G be an upward planar digraph with a vertex u, and let H be an upward planar digraph with a single source u'. Let G' be the graph formed by identifying u and u' in $G \cup H$. Then G' is upward planar. (See Figure 2(b).)*

(c) *Let G be an upward planar digraph with an edge (u, v), and H be an upward planar digraph with a single source u' and a sink v' both on the outer face. Let G' be the graph formed by removing the (u, v) edge of G and adding H, identifying vertex u with u' and vertex v with v'. Then G' is upward planar. (See Figure 2(c).)*

(d) *Let G be a DAG which has an upward planar representation where the cyclic edge order about vertex v is $e_0, \ldots, e_{k-1}$ (vertices $v_0, \ldots v_{k-1}$). Let G' be the DAG formed by splitting v into two vertices: v' incident* *with edges* $e_i, \ldots, e_j$, *and* v'' *incident with edges* $e_{j+1}, \ldots, e_{i-1}$ ($i \neq j$, *arithmetic mod* k). *Then* G' *is upward planar. If* G *had a single source, and* i *and* j *are such that each of* v' *and* v'' *retain at least one incoming edge, then the resulting* G' *is also a single-source digraph.* (*See Figure* 2(*d*).) □

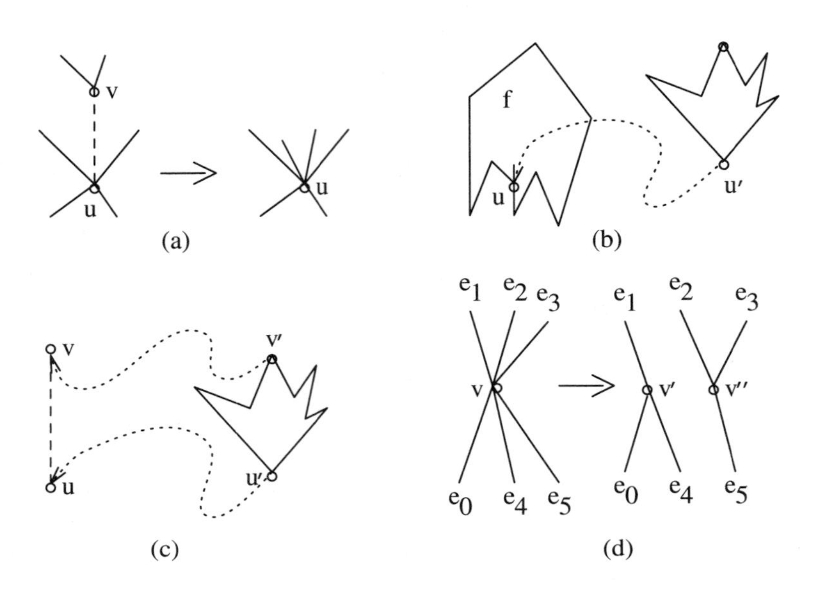

FIGURE 2. Properties of Upward Planar Representation.

Construction (a) generalizes the fact that G is upward planar iff any directed subdivision of G is (cf. [**20**]) by contracting an edge connected to a vertex of in- (out-) degree 1; (b) can be viewed as "inserting" the drawing for H into some face "above" u in the drawing of G; (c) can be viewed as replacing a directed edge in an upward plane drawing of G with another upward plane drawing of H which is, in some sense, "topologically equivalent" to an edge within the drawing of G; (d) allows us to "split" a vertex in two.

6. Separation into Tri-Connected Components

The algorithm of Section 4 tests for upward planarity of a single-source DAG G starting from a given planar representation and outer face of G. In principle, we could apply this test to all planar representations of G, but this would take exponential time. In order to avoid this, we will decompose the graph into biconnected and then into triconnected components. Each triconnected component has a unique planar representation (see [**2**]), and only a linear number of possible outer faces. We can thus test upward planarity of the triconnected components in quadratic time using the algorithm of Section 4. Since we will perform the splitting and merging of triconnected components in quadratic time, the total time will then be quadratic.

To decompose G into biconnected components we use:

LEMMA 6.1. *A DAG G with a single source s and a cut vertex v is upward planar iff each of the k components H_i of G (with respect to v) is upward planar.*

Proof. If G is upward planar then so are its subgraphs the H_i's. For the converse, note that if $v \neq s$ then v is the unique source in all but one of the H_i's; and if $v = s$ then v is the unique source in each H_i. Apply Lemma 5.1(b). □

Dividing G into triconnected components is more complicated, because the cut-set vertices impose restrictive structure on the merged graph. In the biconnected case, it is sufficient to simply test each component separately, since biconnected components do not interact in the combined drawing. The analogous approach for triconnected components would be to add a new edge between the vertices of the cutset in each component, then perform the test recursively. This, however, does not suffice for upward planarity, as illustrated by the two examples in Figure 3. (Recall our convention that direction arrow-heads are assumed to be "upward" unless otherwise specified.) In (a), the union of the graphs is upward planar, but adding the edge (u, v) to each makes the second component non-upward-planar. In (b), the graph is non-upward-planar, but each of the components is upward planar with (u, v) added.

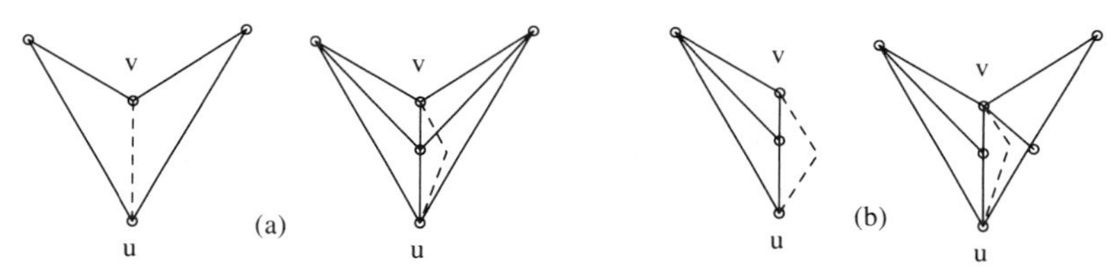

FIGURE 3. Added complication of 2-vertex cut-sets.

We will find it convenient to split the graph G into exactly two pieces at a separation pair $\{u, v\}$, where one of these pieces, E, is a component with respect to the separation pair, and the other piece, F, is the union of the remaining components. This forces each piece to fit into one face of the embedding of the other piece:

LEMMA 6.2. *For G, E, F as above, let Γ be a plane embedding of G, and let Γ_E and Γ_F be the embeddings induced on E and F, respectively. Then in Γ, all of E lies in a single face of Γ_F, and all of F lies in a single face of Γ_E. Furthermore, at least one of E, F, must lie in the outer face of Γ_F, Γ_E, respectively.* □

We will test upward planarity of a biconnected graph G by breaking it at a cut-set into pieces E and F as above, and looking for upward planar embeddings Γ_E and Γ_F that fit together as in Lemma 6.2.

We need a face in Γ_E that contains u and v and is the right "shape" to accommodate the "shape" of Γ_F; and we need a face in Γ_F that contains u and v and is the right "shape" to accomodate the "shape" of Γ_E. (Figure 4(b) showed an example where these conditions fail.) These conditions will be enforced by adding a "marker" connecting u and v to E (F, respectively) that captures the "shape" of Γ_F (Γ_E, respectively), and forces u and v to lie in a common face. For example, the simplest case is when u is the source and v is the sink of F; then the marker representing Γ_F in E is a single (u, v) edge.

Besides playing the primary role described above, the markers will also be used to make the two components 3-connected, and single-source, thus allowing us to recurse on smaller subproblems. The markers we are interested in are shown in Figure 4.

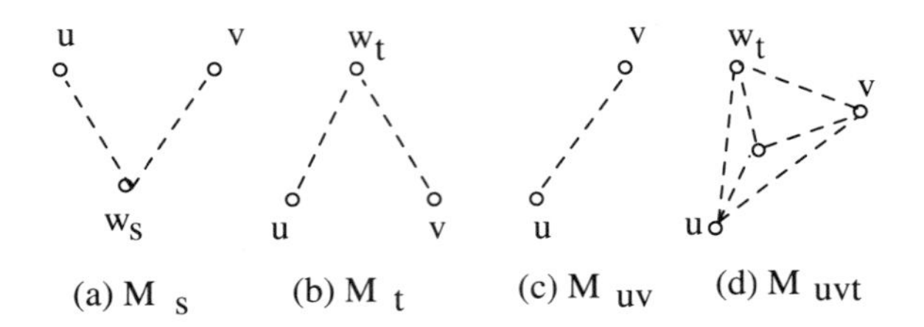

FIGURE 4. Marker Graphs.

We need one other main idea. The last statement of Lemma 6.2 is that one of E, F, must lie in the outer face of Γ_F, Γ_E, respectively. For undirected graphs this causes no problem, since any face can be made the outer one. However, for upward planarity, this condition complicates things. The situation is simplified when $s \neq u, v$. In this case we will take E to be the $\{u, v\}$ component containing s, and so Γ_E must lie in the outer face of Γ_F. When $s \in \{u, v\}$ we must do extra work to decide the "outer" component.

Having determined or decided that Γ_E must lie in the outer face of Γ_F, we know that u and v must be on the outer face of Γ_F. Thus our algorithm will solve the more general problem of testing upward planarity under the condition that some specified set X of vertices, called the "outer" set, must lie on the outer face.

To summarize, given a biconnected graph G and an "outer" set of vertices X, we break G at a cut set $\{u, v\}$ into one component E containing s, and the union of the remaining components F. We add appropriate markers to E and F, specify their "outer" sets, and recurse. We must prove that G has an upward planar embedding with its "outer" set on the outer face iff the smaller graphs do.

The details of this plan make up the remainder of this section. We will consider three cases separately: when u and v are incomparable; when u and v are comparable with $s < u < v$; and when u and v are comparable with $u = s$.

An important note to make at this time is that the markers, except for M_{uv}, are subgraphs attached at only two vertices, which means that $\{u, v\}$ will still constitute a cut-set. For the purposes of determining cut-sets, and making recursive calls, the markers should be treated as distinguished edges—a single edge labelled to indicate its role. As long as the type of marker is identified, the algorithm can continue to treat the vertices of attachment as source, sink or neither, as appropriate for the particular operation.

6.1. Cut-set $\{u, v\}$; u and v are incomparable. Here we consider vertex cut-sets $\{u, v\}$ which are incomparable (then neither is s). We divide the graph G at $\{u, v\}$ into two subgraphs—the *source component* E (the one component which contains the source s), and the union of the remaining components F.

First we need two preliminary results, showing upper and lower bounds (in the partial order corresponding to G) under certain conditions. These allow us to prove the necessity conditions in Theorem 6.1 (to come).

PROPOSITION 6.1. *If G is a connected DAG with exactly two sources u and v, then there exists some w_t such that two vertex disjoint (except at w_t) directed paths $u \overset{+}{\rightarrow} w_t$ and $v \overset{+}{\rightarrow} w_t$ exist in G.* □

LEMMA 6.3. *If G is a biconnected DAG with a single source s, and u and v are incomparable vertices in G, then there exists some w_s such that two vertex disjoint (except at w_s) directed paths $w_s \overset{+}{\rightarrow} u$ and $w_s \overset{+}{\rightarrow} v$ exist in G. If $\{u, v\}$ is a cut-set in G, then there also exists some w_t such that two vertex disjoint (except at w_t) directed paths $u \overset{+}{\rightarrow} w_t$ and $v \overset{+}{\rightarrow} w_t$ exist in G.* □

We are now ready to proceed with the statement of the first main result of the decomposition.

THEOREM 6.1. *Let G be a biconnected directed acyclic graph with a single source s and let $X = \{x_i\} \subseteq V(G)$ be a set of vertices. Let $\{u, v\}$ be a separation pair of G, with u and v incomparable. Let E be the connected component of G with respect to $\{u, v\}$ containing s, and F be the union of all other components. Then, G admits an upward plane drawing with all vertices of X on the outer face if and only if*

(i) *$E' = E \cup M_t$ admits an upward plane drawing with all vertices of X in E on the outer face, and w_t on the outer face if some $x \in X$ is contained in F.*

(ii) *$F' = F \cup M_s$ admits an upward plane drawing with all vertices of X in F on the outer face.* □

Here, as in the remaining cases, the proof has the same basic flavour. The necessity of the marker-conditions follows from the existence of the corresponding marker 'within' (i.e. homeomorphic to a subgraph of) the companion component. The sufficiency is shown by applying the properties of an upward planar representation from Section 4 to combine upward planar representations for the two subproblems into a single upward planar representation; see Figures 5 and 6 for an idea of this process.

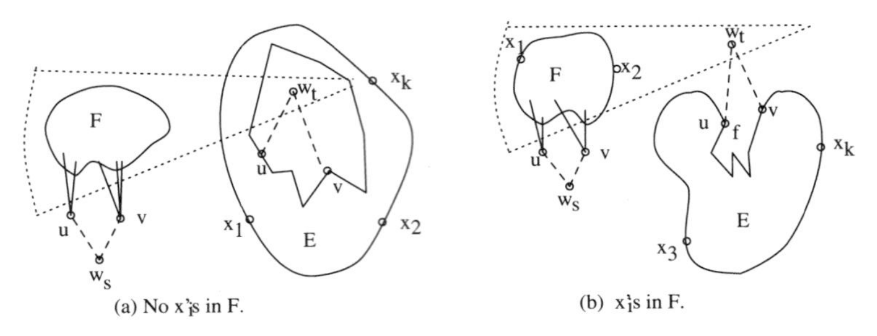

FIGURE 5. Merging E and F; cut-set $\{u, v\}$ incomparable.

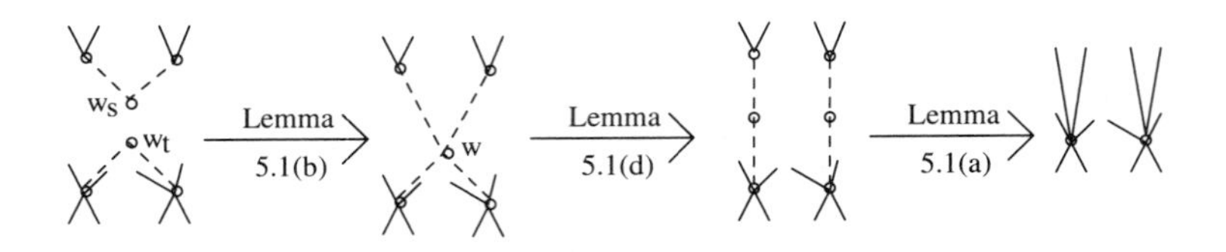

FIGURE 6. Merge construction; $\{u, v\}$ incomparable.

6.2. Cut-set $\{u, v\}$, where $u < v$, $u \neq s$. Here we consider any other vertex cut-sets not involving the source s. We again divide G into the *source component* E and the union of the remaining components F. Note that v can be a source in E, as long as there is a u to v path in F.

An additional preliminary result will be useful. It can be proven by extending the arguments of Proposition 6.1.

LEMMA 6.4. *If G is a biconnected DAG with a single source s and cut-set $\{u, v\}$, where $u < v$ in G and $u \neq s$, then in any non-source component H of G with respect to $\{u, v\}$, where $deg^+ v > 0$, there exists some w_t such that $u \stackrel{+}{\rightarrow} w_t$ and $v \stackrel{+}{\rightarrow} w_t$ are vertex disjoint directed paths in H.* $\square$

We can now continue with the second main result of the decomposition.

THEOREM 6.2. *Let G be a biconnected directed acyclic graph with a single source s, and let $X = \{x_i\} \subseteq V(G)$ be a set of vertices. Let $\{u, v\}$ be a separation pair of G with $u < v$ in G and $u \neq s$. Let E be the source component of G with respect to $\{u, v\}$ and F be the union of all other components. Then, G admits an upward plane drawing with all vertices of X on the outer face if and only if*

(i) *$E' = (E \cup F\text{-marker})$ admits an upward plane drawing with all vertices of X in E on the outer face and w_t (if it exists, otherwise the edge (u, v)) on the outer face if some $x \in X$ is contained in F.*

(ii) $F' = (F \cup E\text{-}marker)$ *admits an upward plane drawing with* w_t *(if it exists, otherwise the edge* (u, v)*) and all vertices of* X *in* F *on the outer face.*

where

$$F\text{-}marker = \begin{cases} M_t & \textit{if } v \textit{ is a source in } F \\ M_{uv} & \textit{if } v \textit{ is a sink in } F \\ M_{uvt} & \textit{otherwise,} \end{cases}$$

and

$$E\text{-}marker = \begin{cases} M_t & \textit{if } v \textit{ is a source in } E \\ M_{uv} & \textit{otherwise.} \end{cases} \qquad \square$$

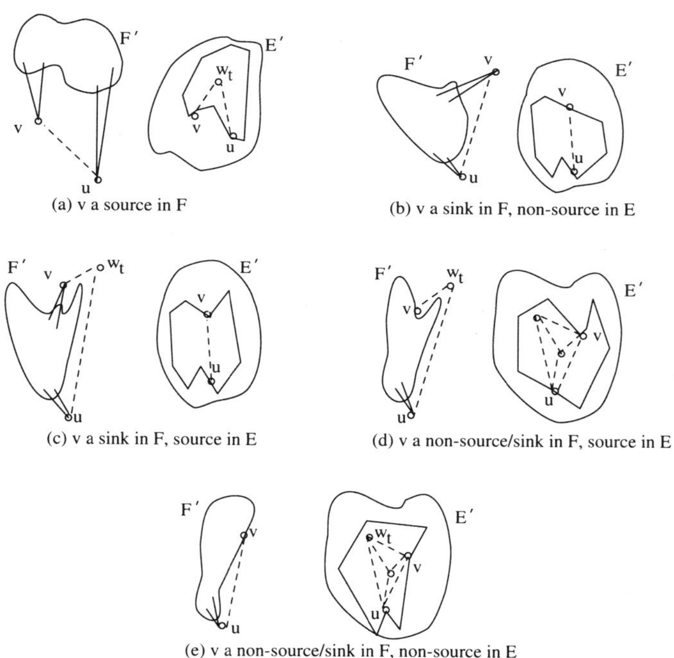

FIGURE 7. Merging E and F; cut-set $\{u, v\}$ and $u < v$.

To prove Theorem 6.2, we apply the same approach as when u and v were incomparable. However, the situation is greatly complicated and we need to use a larger set of markers, and examine more cases. The necessity follows from the existence of the markers as homeomorphic to subgraphs of the components they represent. The sufficiency proof is an extension of the previous method appealing to the new preliminary results of this section. The cases are documented in Figure 7.

6.3. Cut-set $\{s, v\}$. As mentioned in the introduction to Section 6 (see also Lemma 6.2), it is important to be able to distinguish the "inner" and "outer" components. The inner component will be embedded in a face of the outer one, and thus the inner component will have to have the marker on its outer face since this marker is a proxy for the outer component. If we have to check each component as a potential inner component, we must recursively solve two subproblems for each component, and an exponential time blowup results.

Until now, the outer component has been uniquely identified as the *source component*, since that component cannot lie within an internal face of any other component. If we have a cut-set of the form $\{s, v\}$ where s is the source, then we lose this restriction, so we handle it instead by requiring one of the components, E, to be 3-connected so that deciding if it can be the inner face does not require recursive calls. To decide if E can be the inner face we need to test if it satisfies the role of E in the previous theorem—i.e. has an upward planar representation with the marker on its outer face. This can be done in linear time using the algorithm of Section 4. If G has only cut-sets of the form $\{s, v\}$, then, for at least one such cut-set, one of the components will be triconnected. Given the list of cut-sets we can find such a cut-set and such a component in linear time using depth-first search.

We capture these ideas in terms of two theorems. One is applicable if the triconnected component, E, can be the inner component, and one if it cannot. E "can be" the inner component if and only if it satisfies the same conditions that the inner component F satisfied in the previous Theorem 6.2. Both are similar to Theorem 6.2, but a new precondition and some technical detail distinguishes the two cases from it and each other. Note that in the statement of these theorems, we continue to use u (redundant since $u = s$) for consistency with previous usage.

THEOREM 6.3. *Let G be a biconnected DAG with a single source s, and let $X = \{x_i\} \subseteq V(G)$ be a set of vertices. Let $\{u, v\}$ be a separation pair of G where $u = s$, E be a 3-connected component of G with respect to $\{u, v\}$, and F be the union of all other components of G with respect to $\{u, v\}$. If*

(i) *$E' = (E \cup F$-marker$)$ admits an upward plane drawing with w_t (if it exists, otherwise the edge (u, v)) and all vertices of X in E on the outer face,*

then G admits an upward plane drawing with all vertices of X on the outer face if and only if

(ii) *$F' = (F \cup E$-marker$)$ admits an upward plane drawing with all vertices of X in F on the outer face, and w_t (if it exists, otherwise the edge (u, v)) also on the outer face if some $x \in X$ contained in E,*

where

$$E\text{-marker} = \begin{cases} M_t & \text{if } v \text{ is a source in } E \\ M_{uv} & \text{if } v \text{ is a sink in } E \\ M_{uvt} & \text{otherwise,} \end{cases}$$

and

$$F\text{-}marker = \begin{cases} M_t & \text{if } v \text{ is a source in } F \\ M_{uv} & \text{otherwise.} \end{cases} \qquad \square$$

THEOREM 6.4. *Let G be a biconnected DAG with a single source s and let $X = \{x_i\} \subseteq V(G)$ be a set of vertices. Let $\{u, v\}$ be a separation pair of G where $u = s$, E be a 3-connected component of G with respect to $\{u, v\}$, and F be the union of all other components of G with respect to $\{u, v\}$. If it is not true that*

(i) *$E^* = (E \cup F^*\text{-}marker)$ admits an upward plane drawing with w_t (if it exists, otherwise the edge (u, v)) and all vertices of X in E on the outer face,*

where

$$F^*\text{-}marker = \begin{cases} M_t & \text{if } v \text{ is a source in } F \\ M_{uv} & \text{otherwise,} \end{cases}$$

then G admits an upward plane drawing with all vertices of X on the outer face if and only if

(i) *There is no $x \in X$ contained in F.*

(ii) *$F' = (F \cup E\text{-}marker)$ admits an upward plane drawing with w_t (if it exists, otherwise the edge (u, v)) on the outer face.*

(iii) *$E' = (E \cup F\text{-}marker)$ admits an upward plane drawing with all $x \in X$ on the outer face,*

where

$$E\text{-}marker = \begin{cases} M_t & \text{if } v \text{ is a source in } E \\ M_{uv} & \text{otherwise,} \end{cases}$$

and

$$F\text{-}marker = \begin{cases} M_t & \text{if } v \text{ is a source in } F \\ M_{uv} & \text{if } v \text{ is a sink in } F \\ M_{uvt} & \text{otherwise.} \end{cases} \qquad \square$$

7. Conclusions and Further Work

We have given a linear time algorithm to test whether a given single-source digraph has an upward plane drawing strongly equivalent to a given plane drawing, and give a representation for this drawing if it exists. This provides, in combination with the decomposition results of Sections 5 and 6, an efficient $O(n^2)$ algorithm to test upward planarity of an arbitrary single-source digraph. We have also given a combinatorial characterization of single-source upward planar digraphs which provides new insights into their structure.

A lower bound for the single-source upward planarity problem is not known, although we believe that it may be possible to perform the entire test in subquadratic (perhaps linear) time. An obvious extension of this work would be to find such an algorithm or prove a lower bound.

This paper has concentrated on the issues of efficiently testing for an upward plane drawing. However, with the planar representation which results, we can augment the graph to a planar s-t graph with our algorithm of Section 4, then apply the algorithm of DiBattista and Tamassia to generate an actual drawing in $O(n \log n)$ arithmetic steps [**6**]. For actual drawings, the question of what size integer grid (equivalently coordinate precision on a real grid) is required becomes relevant. No upper bound is known on the size of integer grids required for upward planar drawings is known. DiBattista, Tamassia and Tollis [**7**] have shown that there exist upward planar graphs which require an exponential ($\Omega(2^n)$) sized integer grid so any upward-planar drawing algorithm would actually be output sensitive (interestingly different from the case of undirected graphs, which can always be drawn in an $O(n)$ by $O(n)$ integer grid [**4, 17**])—to output $2n$ coordinates of size $\Omega(2^n)$ requires $\Omega(n^2)$ time, which would dominate the arithmetic-steps time. Without an upper bound on the required area, it is not known if the drawing algorithm remains polynomial time (the upper bound on area could be doubly exponential, in which case the output size of the drawing would be itself exponential), but the output is likely impractical even for the singly-exponential grid. Note that this problem disappears if we allow $O(n)$ total bends in the edges [**6, 7**]. It would be interesting to characterize some classes of digraphs which permit straight-line upward plane drawings on a polynomially sized grid. Guaranteeing minimum area in all cases is, however, NP-hard [**8**].

The more general problem of testing upward planarity of an arbitrary acyclic digraph is open. The only known characterization is that any such graph is a subgraph of a planar s-t graph [**6**]. A recent development by Bertolazzi and DiBattista [**1**] shows how to efficiently test a triconnected (multi-source, multi-sink) DAG for upward planarity, a more general analogue of the result in Section 4.

References

1. P. Bertolazzi and G. DiBattista, *On Upward Drawing Testing of Triconnected Digraphs*, Proc. 7th ACM Symposium on Computational Geometry (1991), 272–280. Also TR RAP.18.90 Università À Degli Studi Di Roma, "La Sapienza".
2. J. A. Bondy and U. S. R. Murty, *Graph Theory with Applications*, MacMillian Co. New York, 1976.
3. K. S. Booth and G. S. Lueker, *Testing the consecutive ones property, interval graphs, and graph planarity using PQ-tree algorithms*, J. Comput. Syst. Sci. **13** (1976), 335–379.
4. H. DeFraysseix, J. Pach, and R. Pollack, *Small sets supporting Fary embeddings of planar graphs*, Proc. 20th ACM Symposium on the Theory of Computing (1988), 426–433.
5. G. DiBattista, W. Liu and I. Rival, *Bipartite graphs, upward drawings, and planarity*, Inf. Proc. Letters **36** (1990), 317–322.
6. G. DiBattista and R. Tamassia, *Algorithms for plane representations of acyclic digraphs*, Theoretical Computer Science **61** (1988), 175–178.
7. G. DiBattista, R. Tamassia and I. G. Tollis, *Area requirement and symmetry display in drawing graphs*, Discrete and Computational Geometry **4** (1992), 381–401.
8. D. Dolev, F. T. Leighton and H. Trickey, *Planar embedding of planar graphs*, in *Advances in Computing Research, Vol. 2*, F. P. Preparata, ed., JAI Press Inc., Greenwich, Connecticut, 147–161.
9. I. Fáry, *On straight line representations of planar graphs*, Acta. Sci. Math. Szeged **11**

(1948), 229–233.

10. I. S. Filotti, G. L. Miller, J. Reif, *On determining the genus of a graph in $O(v^{O(g)})$ steps*, Proc. 11th ACM Symposium on the Theory of Computing (1979), 27–37.
11. J. Hopcroft and R. Tarjan, *Efficient planarity testing*, J. ACM **21** (1974) 549–568.
12. J. Hopcroft and R. Tarjan, *Dividing a graph into triconnected components*, SIAM J. Comput. **2** (1972), 135–158.
13. M. D. Hutton and A. Lubiw, *Upward planar drawing of single source acyclic digraphs*, Proc. 2nd ACM/SIAM Symposium on Discrete Algorithms (1991), 203–211.
14. D. Kelly, *Fundamentals of planar ordered sets*, Discrete Math **63** (1987), 197–216.
15. D. Kelly and I. Rival, *Planar lattices*, Can. J. Math. **27** (1975), 636–665.
16. C. R. Platt, *Planar lattices and planar graphs*, J. Comb. Theory (B) **21** (1976), 30–39.
17. W. Schnyder, *Embedding planar graphs on the grid*, Proc. 1st ACM/SIAM Symposium on Discrete Algorithms (1990), 138–148.
18. R. E. Tarjan, *Depth-first search and linear graph algorithms*, SIAM J. Comput. **1** (1972), 146–159.
19. R. Tamassia and P. Eades, *Algorithms for drawing graphs: an annotated bibliography*, Brown University TR CS-89-09, 1989.
20. C. Thomassen, *Planar acyclic oriented graphs*, Order **5** (1989), 349–361.
21. W. T. Tutte, *Convex representations of graphs*, Proc. London Math. Soc. **10** (1960), 304–320.

DEPARTMENT OF COMPUTER SCIENCE, UNIVERSITY OF TORONTO, TORONTO, ONTARIO M5S 1A4, CANADA
E-mail address: mdhutton@cs.utoronto.ca

COMPUTER SCIENCE DEPARTMENT, UNIVERSITY OF WATERLOO, WATERLOO, ONTARIO N2L 3G1, CANADA
E-mail address: alubiw@maytag.uwaterloo.ca

DIMACS Series in Discrete Mathematics
and Theoretical Computer Science
Volume 9, 1993

Flow in Planar Graphs: A Survey of Recent Results

SAMIR KHULLER AND JOSEPH (SEFFI) NAOR

September 23, 1992

ABSTRACT. We briefly review some of the recent results on flow in planar graphs. The older results in the area are reviewed somewhat more briefly. The main interest in planar flow was regenerated when it was observed in both [**41, 34**] that the single source/sink edge capacity flow problem was not the "basic" problem that should be studied. This survey paper essentially reviews some of the techniques that were developed to capture more general versions of the planar flow problem. Some other interesting results related to planar flow problems are also reported.

1. Introduction

The computation of a maximum flow in a graph has been an important and well studied problem, both in the fields of Computer Science and Operations Research. Many efficient algorithms have been developed to solve this problem, see e.g., [**13**]. Research on flow in planar graphs is motivated by the fact that more efficient algorithms, both sequential and parallel, can be developed by exploiting the planarity of the graph. This is important, in particular for parallel algorithms, since maximum flow in general graphs was shown to be P-complete [**15**]. The planar flow algorithms are not only "good" because they are extremely efficient, but they are also very *elegant.* Planar networks also arise in practical contexts such as VLSI design and communication networks; therefore, it is of interest to find fast flow algorithms for this class of graphs.

1991 *Mathematics Subject Classification.* Primary 68R10, 05C38; Secondary 90B10, 90C35, 90C27.

Key words and phrases. planar graphs, network flow, min-cut, circulation.

The first author was supported in part by NSF Grants CCR-8906949, CCR-9103135 and CCR-9111348.

This paper is in final form, and no version of it will be submitted for publication elsewhere.

In the popular formulation of the *planar* flow problem, one considers single source and sink vertices, s and t. Each edge has a capacity, and one wishes to find the max-flow from s to t. This problem has been extensively investigated by many researchers starting from the pioneering work by Ford and Fulkerson [**7**] who suggested an efficient way for computing the minimum cut in the special case of st-graphs (when the source and sink are on the same face). Berge and Ghouila-Houri [**1**] later developed an $O(n^2)$ time algorithm for computing the flow function in this case. This algorithm was later improved to an $O(n \log n)$ time algorithm by [**26**]. By introducing the concept of potentials, Hassin [**21**] gave an elegant algorithm that can be implemented in $O(n\sqrt{\log n})$ time using Frederickson's shortest path algorithm [**8**]. Itai and Shiloach [**26**] also developed an algorithm to find a max flow in an undirected planar graph when the source and sink are not on the same face. Reif [**49**] showed how to find the minimum cut in this case in $O(n \log^2 n)$ time. Hassin and Johnson [**22**] finally completed the picture by giving an $O(n \log^2 n)$ algorithm to compute the flow function as well. Frederickson speeded up both these algorithms by an $O(\log n)$ factor by giving faster shortest path algorithms [**8**]. The problem of finding a minimum cut in a directed planar graph turned out to be much harder and was first solved by [**29**] (both sequentially and in parallel).

Miller and Naor [**41**] pointed out that the general maximum flow problem in planar graphs is when there are many sources and sinks. Note that one cannot reduce the multiple source-sink problem to the single source-sink version since the reduction may destroy planarity. However, we would like to design more efficient algorithms for this case: for sequential algorithms, we would like to take advantage of the planarity and improve on the performance of the best algorithms for general graphs; in parallel, it is not even known whether an NC algorithm exists for this problem. Miller and Naor [**41**] showed that when demands and supplies are fixed, the problem can be reduced to a "circulation problem" (with lower bounds on edge capacities), and also gave an efficient algorithm for this case. They also gave an efficient algorithm for the case where the demands and supplies are variable, but the sources and sinks belong to a bounded number of faces. If the sources and sinks belong to an arbitrary number of faces, computing a maximum flow efficiently is still open.

Khuller and Naor [**34**] considered the more general flow problem in which vertices as well as edges have capacity constraints. Vertex capacities may arise in various contexts such as computing vertex disjoint paths in graphs [**36**], and in various network situations when the vertices denote switches and have an upper bound on their capacities. For the case of general graphs this problem can be reduced to the version with only edges having capacity constraints by a simple idea of "splitting" vertices into two and forcing all the flow to pass through a "bottleneck" edge in-between. In planar graphs, this reduction may *destroy* the planarity of the graph and thus cannot be used. (The reduction is described in Bondy and Murty ([**4**], page 205) from which the violation of planarity is obvious.)

Notice that in the case of general graphs, as opposed to planar graphs, the single source-sink problem with edge capacities is usually the "basic" problem, since most other formulations of the flow problem can be easily reduced to this problem. It is not clear if there is such a "basic" problem in the context of flow in planar graphs.

An application where vertex capacities play an important role is in reconfiguring VLSI/WSI (Wafer Scale Integration) arrays. Assume that the processors on a wafer are configured in the form of a grid, and due to yield problems, some are going to be faulty. Instead of treating the whole wafer as defective, the non-faulty processors can be reconfigured in the form of a grid. We assume that multiple data tracks are allowed along every grid line. It was shown in [**51**] that in this context, the reconfiguration problem can be abstracted combinatorially as finding a set of *vertex disjoint paths* from the faulty processors (the sources) to the boundary of the grid (the sink). This is a special case of a multiple source/single sink planar flow problem where all vertex capacities are equal to 1. This problem is also referred to as the *escape problem* in the textbook by Cormen, Leiserson and Rivest [CLR, page 626]. The algorithm given by [**51**] has a running time of $O(n^2 \log n)$ where n is the number of grid points. The algorithms of [**34**] improve over this result by an $O(\sqrt{n})$ factor. The reader is referred to [**3, 6, 18, 50, 51**] for more details and bibliography of this problem and on the connection between flow problems and reconfiguration. (The main concern of [**50, 3**] is the single-track model.)

The problem of computing a maximum flow is closely related to the problem of computing a maximum matching. In fact, in a bipartite graph, computing a maximum matching is a special case of computing a maximum flow. In the parallel context, for arbitrary graphs, only a randomized procedure is known for computing a maximum matching [**32, 42**] and also for determining whether a graph contains a perfect matching. In contrast, in planar graphs, it can be decided deterministically in NC [**38, 54**] whether a graph has a perfect matching. (Even the number of perfect matchings can be counted in NC!) However, it is an open question whether a perfect matching can be computed in NC. It follows from the results of [**41**] that a perfect matching in planar bipartite graphs can be computed in NC.

We now move to a slightly different perspective on the planar flow problem. We study the *set of integer solutions* to the planar circulation problem, and characterize an encoding for all the feasible integer circulations. It turns out that the set of feasible circulations in planar graphs forms a distributive lattice where the meet and join operations are defined appropriately. Other examples of problems where the solution set has a similar structure are the stable marriage problem, and the minimum cut problem. Picard and Queyranne [**45**] have shown that the set of all minimum st-cuts forms a distributive lattice where the join and meet operations are defined as intersection and union respectively. The structure of the solution set of the stable marriage problem has been extensively investigated in the book by Gusfield and Irving [**19**].

A brief outline of the paper is as follows: In Section 2 we discuss some basic flow notation used in the rest of the paper. Section 3 is a survey of the past research on planar flow. Section 4 describes the results of [**41**]. In Section 5 we describe the lattice structure of planar flow [**35**]. Section 6 describes the results of [**34**]. Section 7 outlines some related flow problems that are NP-complete even for planar graphs [**2**]. We conclude by outlining some of the major open problems in Section 8.

2. Terminology and preliminaries

We are going to assume that the graph $G = (V, E)$ has a fixed planar embedding. For each edge $e \in E$, let $D(e)$ be the corresponding *dual edge* connecting the two faces bordering e. Let $\mathcal{D} = (F, D(E))$ be the *dual graph* of G, where F is the set of faces of G and $D(E) = \{D(e)|e \in E\}$. There is a 1-1 correspondence between primal and dual edges and the direction of a primal edge e induces a direction on $D(e)$. We use a left hand rule: if the thumb points in the direction of e, then the index finger points in the direction of $D(e)$ (keeping the palm face up). For a vertex v, $in(v)$ refers to the arcs that are carrying incoming flow to vertex v. Similarly $out(v)$ refers to those arcs that are carrying flow out of the vertex v.

Associate with each edge $e \in E$, a capacity $c(e) \geq 0$, and also with each vertex $v \in V - \{S, T\}$, a capacity $c(v) \geq 0$. Let $S = s_1, \ldots, s_l$ and $T = t_1, \ldots, t_k$ be two sets of distinguished vertices, called *sources* and *sinks* respectively. We assume that the vertices in S and T have no capacities. Otherwise, suppose that vertex $s \in S$ has capacity $c(s)$; add to the graph a new distinguished vertex s' adjacent only to s, such that the capacity of the edge joining s and s' is unbounded. Remove vertex s from S and add s' to S. By performing this step for every capacitated vertex in S, and an analogous step for every capacitated vertex in T we obtain the required property.

A function $f : E \rightarrow Z$ is a legal flow function if and only if:

(i): $\forall e \in E : 0 \leq f(e) \leq c(e)$.

(ii): $\forall v \in V - \{S, T\} : \sum_{e \in in(v)} f(e) = \sum_{e \in out(v)} f(e)$.

(iii): $\forall v \in V - \{S, T\} : \sum_{e \in in(v)} f(e) \leq c(v)$.

We assume that G is biconnected; otherwise, we can add edges with zero capacities appropriately to ensure that.

The *cost* of a dual edge is defined in the undirected case to be the capacity of the corresponding primal edge. (The dual edge is also undirected.) In the directed case, given a primal edge e of capacity $c(e)$, it has two dual edges corresponding to it: one is directed according to the left hand rule and has cost $c(e)$; the other is in the converse direction and has cost 0, or in general, its cost is equal to the lower bound on the flow on edge e (see Fig. 1).

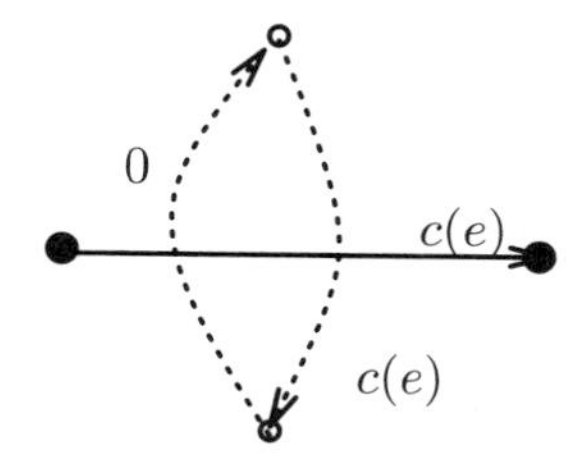

FIGURE 1. Dual graph in case of a directed graph

In the maximum flow problem, we are looking for a legal flow function that maximizes the amount of flow entering T (or leaving S). The amount of flow entering the sink is also called the *value* of the flow function. A *circulation* is a legal flow function where condition (ii) is applied to every vertex in the graph, i.e., there are no sources and sinks.

A natural generalization of the flow problem is when edges have a lower bound different from zero on their capacity; in this case, the capacity of an edge will be denoted by $[a, b]$, where $a \leq b$.

The *residual graph* is defined with respect to a given flow. Let $e = (v, w)$ be an edge with capacity $[a, b]$ and flow f. In the residual graph e is replaced by two directed edges (v, w) and (w, v) with capacities $[0, b - f]$ and $[0, f - a]$ respectively.

A *spurious cycle* is a directed cycle along which the flow can be reduced, without any of the edges violating the lower bounds on their capacities.

A special case of planar flow is when the source and sink are on the same face. These graphs are called *st*-graphs.

A *potential* function $p : F \to Z$ is defined on the faces of a planar graph. Let e be an edge in the graph G, and let $D(e) = (g, h)$ be its corresponding edge in the dual graph such that $D(e)$ is directed from g to h. The potential difference over e is defined to be $p(h) - p(g)$. The following proposition, proved in [Ha] and [Jo], can be easily verified.

PROPOSITION 2.0.1. *Let $C = c_1, \ldots, c_k$ be a cycle in the dual graph and let $f_1, \ldots, f_k$ be the potential differences over the cycle edges. Then,* $\sum_{i=1}^{k} f_i(e) = 0$.

It follows from the proposition that the sum of the potential differences over all the edges adjacent to a primal vertex is zero.

A potential function is defined to be *edge consistent* if the potential difference over each edge is not larger than its capacity. Such a potential function induces a circulation in the graph. If the circulation satisfies the vertex capacities as well, the potential function is defined to be *consistent.* The use of a potential function as a means of computing a flow was first suggested by Hassin [**21**], and was later elaborated by [**22**] and [**29**]. Miller and Naor [**41**] use the idea of potentials to solve the problem of computing a feasible circulation as well.

The model of parallel computation used is the Exclusive-Read Exclusive-Write (EREW) Parallel Random Access Machine (PRAM). A PRAM employs synchronous processors all having access to a shared memory. An EREW PRAM does not allow simultaneous access by more than one processor to the same memory location.

3. Planar flow with a single source and sink

All the results referred to in this section deal *exclusively* with the single source, single sink maximum flow problem where only edges have capacities. Ford and Fulkerson [**7**] had already observed that a minimum cut in a planar graph is equivalent to a minimum weight cycle that separates the source from the sink in the dual graph. They gave an $O(n \log n)$ time algorithm to compute the minimum cut when the source and sink belong to the same face. Berge and Ghouila-Houri [**1**] suggested an $O(n^2)$ algorithm for computing the flow function which is called the "uppermost path algorithm". This algorithm was implemented in $O(n \log n)$ time by Itai and Shiloach [**26**]. Hassin [**21**] gave an elegant algorithm to compute the flow function and his algorithm can be implemented in $O(n\sqrt{\log n})$ time using the method of [**8**] for computing shortest paths in planar graphs.

Hassin's idea can be summarized as follows: Partition the infinite face (where the source and sink are assumed to be) by a directed edge (from t to s) of infinite capacity. Let s^* denote one of the two faces generated by this partitioning as in Fig. 4 (the return edge is not shown in the picture). For each edge $(i, j) \in E$, let $(i', j') \in D(E)$ be the associated dual edge.

DEFINITION 3.0.1. *The potential $p(f)$ of a face f (or a vertex f in the dual graph) is defined to be the length of the shortest distance from s^* to f.*

The flow on edge (i, j) is defined as follows: $f(i, j) = p(j') - p(i')$. This yields an edge consistent potential function since $p(j') \leq c(i, j) + p(i')$ (the potentials satisfy the shortest path property), and hence a valid flow function. The flow moves through the edge always keeping the face with higher potential to its right.

Itai and Shiloach [**26**] also gave an algorithm to compute the maximum flow in an undirected graph when the source and sink do not necessarily belong to the same face. The main idea of their algorithm is the following: find a path from the source to the sink; send the flow from the source to the sink edge-by-edge. The problem of sending the flow on each edge is an instance of an st-graph problem. (The tail and head of the edge are the source and sink respectively.) Notice that if too much flow is sent initially to the sink, some of it can be "returned" to the source by the same method. The running time of the algorithm is $O(n^2 \log n)$.

Reif [**49**] gave an improved algorithm for computing the minimum cut in this case, i.e., an undirected planar graph where the source and sink are not on the same face. The running time of this algorithm was $O(n \log^2 n)$ time. Hassin and Johnson [**22**] completed the picture by giving an $O(n \log^2 n)$ time algorithm to compute the flow function as well, by generalizing the ideas of [**21**] and [**49**]. (The running time of their algorithm can be improved by using the methods of [**8**] for computing shortest paths in planar graphs.)

It is easy to obtain a parallel implementation of the algorithm given by [**22**] for undirected graphs and its complexity is $O(\log^2 n)$ time using $O(n^3)$ processors. (The details are given in [**29**].) A more efficient implementation that uses only $O(n^{1.5})$ processors can be obtained by using the methods of [**46, 47, 48**]. An alternative algorithm for computing the minimum cut in parallel in an undirected graph was given by [**30**].

The problem of computing a maximum flow in a directed planar graph (when the source and sink are not on the same face) turned out to be more difficult. The intuitive reason for this difficulty is the following. Finding the minimum cut is equivalent to finding a minimum weight cycle in the dual graph separating the source from the sink. In the undirected case, this problem can be reduced to the problem of computing a (certain) minimum weight path. This reduction cannot be applied in the directed case. This problem was eventually solved by Johnson [**29**] who provided both sequential and parallel algorithms for finding the minimum cut as well as the flow function. The complexity of the sequential algorithm is $O(n^{1.5} \log n)$ time (see also [**31**]). The parallel complexity is $O(\log^3 n)$ time using $O(n^4)$ processors, or $O(\log^2 n)$ time using $O(n^6)$ processors. Again, a more efficient implementation can be obtained by using the methods of [**46, 47, 48**].

In the course of the evolution of efficient algorithms for planar flow, an interesting phenomenon occurred. The computational difficulty alternated between searching for the minimum cut on one hand, and computing the flow function, when the minimum cut is known, on the other hand.

4. Planar flow with multiple sources and sinks

The potential method pioneered by Hassin has really paved the way for the future planar flow algorithms (which are clever elaborations of Hassin's basic potential method). A very elegant scheme for computing a flow for the multiple source/sink (when the sources and sinks have fixed supplies and demands) problem was given by Miller and Naor [**41**]. We proceed to outline the scheme in this section.

We then address the problem of computing the maximum flow in the case where the demands and supplies are variable, but the sources and sinks belong to a bounded number of faces. The problem when the demands and supplies belong to an arbitrary number of faces, is open. For sequential algorithms, we would like to take advantage of the planarity and improve on the performance of the best algorithms for general graphs; in parallel, we would like to provide NC algorithm for this problem. Unfortunately, this problem is still open.

4.1. The potential method. In this section we assume that the supply at each source and the demand at each sink is known, and give an efficient algorithm that computes the flow function in this case. We denote the supply of source s_i, $1 \leq i \leq l$, by $|s_i|$ and the demand at sink t_j, $1 \leq j \leq k$, by $|t_j|$. The key idea is to compute a potential function on the faces of the planar graph such that the flow on each edge is the potential difference of the two faces that border the edge. To achieve this, we reformulate the problem as a circulation problem with lower bounds. This is done by first computing a spanning tree T (the orientation of the edges of the graph is ignored for the purpose of computing T). Then, new edges, parallel to the edges of T, are added to the graph to redirect the flow from the sinks back to the sources.

An edge $e \in T$ separates the tree into two parts, called right and left, where T_r, the right part of the tree. Let w_e be $\sum_{t_i \in T_r} |t_i| - \sum_{s_i \in T_r} |s_i|$. A new edge e' parallel to e is inserted with capacity $[w_e, w_e]$ and is directed from T_r to T_l. (It returns the flow from T_r to T_l.) Assigning a lower bound which is equal to the upper bound forces the flow on e' to be equal to w_e. This construction is repeated for each $e \in T$ in parallel and the new graph that results is denoted by G'.

We claim that there is a 1-1 correspondence between flows satisfying the supplies and demands in G and circulations in G'. A circulation G' is computed as follows: pick an arbitrary face in the dual of G' as a root, and compute all shortest paths from it; the distance of a face u from the root, $d(u)$, is defined to be the potential function.

Sketch of Algorithm:

Step 1. In the graph G, compute a spanning tree T. (The orientation of G is ignored in the computation of the spanning tree.)

Step 2. For each edge $e \in T$, compute its return flow: it is equal to the flux between the two parts of the tree which is $w_e = \sum_{t_i \in T_r} |t_i| - \sum_{s_i \in T_r} |s_i|$.

Step 3. For each edge $e \in T$: adjust its weight in G' by adding $[w_e, w_e]$ to its weight. Let $\mathcal{D}' = (F', D(E'))$ be the dual graph of G'.

Step 4. Pick an arbitrary face in F' and compute all shortest distances from it in $\mathcal{D}'$.

Step 5. $\forall v \in F' : p(v) \leftarrow d(v)$

Step 6. $\forall e \in E : f(e) \leftarrow (p(v) - p(u))$ where v and u are the faces that border e and $D(e)$ is oriented from u to v. Now delete all the edges e' that carry "return" flow.

We have to show that the flow function that we compute is both legal and satisfies the demands and supplies. In the following it is assumed that there exists a flow function that satisfies the given demands and supplies.

THEOREM 4.1.1. *The algorithm computes a feasible flow satisfying the demands and supplies in the graph.*

PROOF. It is evident from the construction that every circulation in G' induces a flow f satisfying the demands and supplies in G, and a circulation in G' exists if a feasible flow exists in G. Hence it suffices to compute a circulation in G'. We claim that the existence of a circulation in G' implies the existence of a consistent potential function in $\mathcal{D}'$, the dual graph of G'. Clearly, a consistent potential function p exists if and only if there are no negative weight cycles in $\mathcal{D}'$. By the correspondence between cycles and cuts in planar graphs, a negative cycle in $\mathcal{D}'$ implies that there exists a cut in G', where the lower bounds on the edges leaving one portion of the graph add up to more than the upper bounds of the edges entering that portion. This immediately violates the existence of a feasible circulation in G'. Since there cannot be cycles with a negative weight in $\mathcal{D}'$, a shortest path labeling from an arbitrary face is a consistent potential function that induces a circulation. □

THEOREM 4.1.2. *The sequential running time of the algorithm is $O(n^{1.5})$; the parallel running time is $O(\log^3 n)$ in the EREW PRAM model and $O(\log n)$ time in the CRCW PRAM model, where the number of processors is $O(n^{1.5})$.*

PROOF. The most expensive part of the algorithm is Step (4) where all shortest paths from a vertex in the dual graph are computed. The sequential complexity of computing all shortest paths from a given vertex in a graph with negative edge weights is $O(n^{1.5})$ using the generalized nested dissection of [**39**]. To compute shortest distances in parallel, the nested dissection algorithm of [**46, 47, 48**] requires $O(n^{1.5})$ processors; the time complexity is $O(I(n) \log n)$ where $I(n)$ is the parallel time of computing the sum of n values. $I(n)$ can be implemented in $O(\log n)$ time on the EREW PRAM model. To implement the method of nested dissection we need to compute small separators in planar graphs. A small separator can be computed sequentially in linear time, see [**43**]. In parallel, Gazit and Miller [**11, 12**] provided a procedure for computing small separators that uses $O(n^{1+\epsilon})$ processors and $O(\log^3 n)$ time. (The running time is dominated by the procedure for computing a maximal independent set in a graph. The current best bound is $O(\log^3 n)$ time using $O(m/\log n)$ processors [**14**].)

An embedding of a planar graph can be computed sequentially in linear time [**23**], and in parallel, in $O(\log^2 n)$ time using a linear number of processors [**37**]. The embedding is needed for computing the dual graph. □

4.2. Maximum flow for a bounded number of faces containing sources and sinks. In this section we provide efficient sequential and parallel algorithms for the case when the sources and sinks belong to a fixed number of faces. Without loss of generality, we can assume that all capacities are nonnegative and that the sources and sinks alternate on every face, namely there are no two consecutive sources (or sinks). This property will be maintained during the recursive calls to the algorithm.

Let G be a graph and f a maximum flow. The *Ford-Fulkerson cut* with respect to f is the set of edges between W, the vertices reachable by an augmenting path in $G - f$, and $V - W$. That is, the Ford-Fulkerson cut is the "first" minimum cut separating the sources from the sinks.

We first claim that the following generic algorithm computes a maximum flow in a graph G, (not necessarily planar), with many sources and sinks.

(i) Partition the sources and sinks into two disjoint sets L and R.
(ii) Compute the maximum flow from L to R, i.e., from the sources in L to the sinks in R. Let C denote the Ford Fulkerson minimum cut with respect to the maximum flow.
(iii) Remove the edges of C from the graph G. Compute recursively a maximum flow in each connected component.
(iv) Compute a maximum flow from R to L.

Notice that the input graph in Steps (3) and (4) is the residual graph.

Let G be a planar graph with variable sources and sinks that lie on at most k faces of G, denoted by $F_1, \dots F_k$. We now show how to implement the generic algorithm to compute a maximum flow in G efficiently.

Step (1) in the generic algorithm is implemented as follows: the sources and sinks on face F_i $(1 \leq i \leq k)$ are partitioned into two sets, L_i and R_i, such that the set L_i contains as many sources as the set R_i and both sets are approximately of the same cardinality. The set L is defined to be the union of the sets L_i $(1 \leq i \leq k)$ and the set R is defined to be the union of the sets R_i $(1 \leq i \leq k)$. The maximum flow from L to R is defined to be the flow that maximizes the flow from the sources in L to the sinks in R. In particular, we can set the demand of each sink in L and each source in R to zero.

To implement Step (2), first connect the sources in each set L_i to a super source s_i, and the sinks in each set R_i to a super sink t_i. This operation does not violate the planarity of the graph. Observe that the following greedy algorithm computes a maximum flow with many sources and sinks: for all pairs i, j, $1 \leq i, j \leq k$, compute in arbitrary order the maximum flow from s_i to t_j; the flow is computed in the residual graph with respect to the pairs for which a maximum flow has already been computed. Computing the flow from source s_i to sink t_j is an instance of the problem of computing the maximum flow in a directed planar graph with a single source and single sink. Step (4) is implemented similarly. We refer the reader to [**41**] for the proof of this algorithm.

Notice that in Step (3), the number of alternations of sources and sinks is reduced by a constant factor for each face F_i $(1 \leq i \leq k)$.

We conclude with the following theorem.

THEOREM 4.2.1. *If G is a planar flow graph with variable sources and sinks* *that lie on at most k faces of G, then a maximum flow for G can be computed (sequentially and in parallel) in $O(k^2)$ calls to a procedure that computes a maximum flow in the single source and single sink case.*

5. The lattice structure of planar flow

In this section we study the *structure* of the set of solutions to the circulation problem for planar graphs. We first establish a one-to-one correspondence between consistent potential functions and circulations. Given a legal circulation C, a corresponding potential function can be constructed as in the proof of Theorem 4.1.1. (Pick an arbitrary face r and compute all shortest distances from it where the edge weights in this search are the actual flows in the circulation.) It is not hard to see that this potential function induces circulation C. We shall henceforth view a potential function as a vector where the entries correspond to the potentials of the faces, and the potential of the *root face* (an arbitrary but fixed face) is always equal to zero.

Given two consistent vectors, P_1 and P_2, we say that $P_1 \geq P_2$ if for all components i, $P_1(i) \geq P_2(i)$. We say that circulation C_1 *dominates* C_2 if, for their corresponding potential vectors P_1 and P_2, $P_1 \geq P_2$. We use the term $\mathcal{P}$ to refer to set of all the consistent potential vectors. It is easy to see that $\mathcal{P}$ is a partial order under the dominance relation (also written as $(\mathcal{P}, \preceq)$). We claim that the set $\mathcal{P}$ very naturally forms a distributive lattice, where a distributive lattice is a partial order in which:

(i) Each pair of elements has a greatest lower bound, or *meet*, denoted by $a \wedge b$, so that $a \wedge b \preceq a, a \wedge b \preceq b$, and there is no element c such that $c \preceq a, c \preceq b$ and $a \wedge b \prec c$.

(ii) Each pair of elements has a least upper bound, or *join*, denoted by $a \vee b$, so that $a \preceq a \vee b, b \preceq a \vee b$, and there is no element c such that $a \preceq c, b \preceq c$ and $c \prec a \vee b$.

(iii) The *distributive* laws hold, namely $a \vee (b \wedge c) = (a \vee b) \wedge (a \vee c)$ and $a \wedge (b \vee c) = (a \wedge b) \vee (a \wedge c)$.

Given two circulations C_1 and C_2 (represented as P_1 and P_2), we define the meet as the circulation induced by the potential vector $P_m = \min(P_1, P_2)$. Clearly, the face at zero potential in both circulations stays at zero potential. Every face g is assigned a potential equal to $\min(P_1(g), P_2(g))$ where $P_i(g)$ is the potential of g in C_i. Similarly, the join is defined as $P_j = \max(P_1, P_2)$. The proof of the following theorem is not hard.

THEOREM 5.0.1. *The partial order $(\mathcal{P}, \preceq)$ is a distributive lattice, with the meet and join defined appropriately.*

It is easy to see that a lattice has a unique minimum and maximum, P_b and P_t, referred to as *bottom* and *top* respectively. We now provide a simple characterization for them. The shortest path potential vector, in which the potential of a face is exactly its distance from the root face in the dual graph corresponds precisely to the top of the lattice. A vector corresponding to the bottom of the lattice can be computed as follows: the potential of face f is the length of the shortest path from f to r, the root face, multiplied by -1.

It turns out that the flow functions computed by **[21]**, **[22]**, **[29]** and **[41]** can be interpreted as corresponding to the top element of the circulation lattice. (This is essentially a matter of notation; if we reverse the direction of the dual edges, their algorithms will be computing the bottom element in the lattice.)

The lattice representing all feasible circulations is clearly of *exponential* size, since there are exponentially many solutions to the circulation problem. A compact encoding of the entire lattice can be obtained by constructing a directed acyclic graph such that the predecessor-closed subsets of this partial order correspond to elements in the lattice. Although this DAG may be large, its size depends on the maximum edge-capacity – it can be represented succinctly, in polynomial size. This compact encoding of the partial order provides in turn a compact encoding of the lattice elements.

There is also a connection between the lattice and unidirectional cycles. Recall that we assumed the planar embedding was such that the infinite face is the root face. Each simple cycle divides the sphere into two nonempty disjoint sets of faces, called regions. The region containing the root face is designated the *exterior* region; the other region is *interior*. In a traversal of a directed cycle, all faces that border the cycle on its right are in the same region, the cycles *right-hand region*.

DEFINITION 5.0.2. *A directed cycle is* clockwise *if the cycles right-hand region is interior. Otherwise, the cycle is* counterclockwise.

The top and bottom of the lattice can be characterized by unidirectional cycles.

THEOREM 5.0.3. *A circulation is clockwise maximal if and only if it corresponds to P_t. A circulation is counterclockwise maximal if and only if it is P_b.*

It is tempting to believe that the dominance relation in the lattice can be stated in terms of saturating clockwise cycles. That is, if $P_1 < P_2$, then circulation P_2 can be obtained from P_1 by saturating clockwise cycles. Unfortunately, the following counterexample shows that this is not true. Let c_1 and c_2 be clockwise cycles such that c_1 is contained in the interior of c_2. We construct two circulations, P_1 and P_2, such that $P_1 < P_2$. To construct P_1, take P_b and push one unit of flow in the cycle c_1 in the clockwise direction. To construct P_2, take P_b and push one unit of flow in the cycle c_2 in the clockwise direction. Obviously, $P_1 < P_2$, but the only way to obtain circulation P_2 from P_1 is to push a unit of flow in the cycle c_2 in the clockwise direction and in the cycle c_1 in the counterclockwise direction.

6. Planar flow with vertex capacities

In this section we survey some of the results in [**34**] that address the more general problem when vertices as well as edges have capacities. As we already have mentioned, the standard trick in general graphs for eliminating vertex capacities may *destroy* the planarity of the graph.

Assume that a planar flow problem with vertex capacities is reduced to a circulation problem with vertex capacities. It is interesting to note that in this case, the set of feasible circulations does *not* form a lattice. This observation may partly account for the computational difficulty in computing flows in the case where the vertices are capacitated.

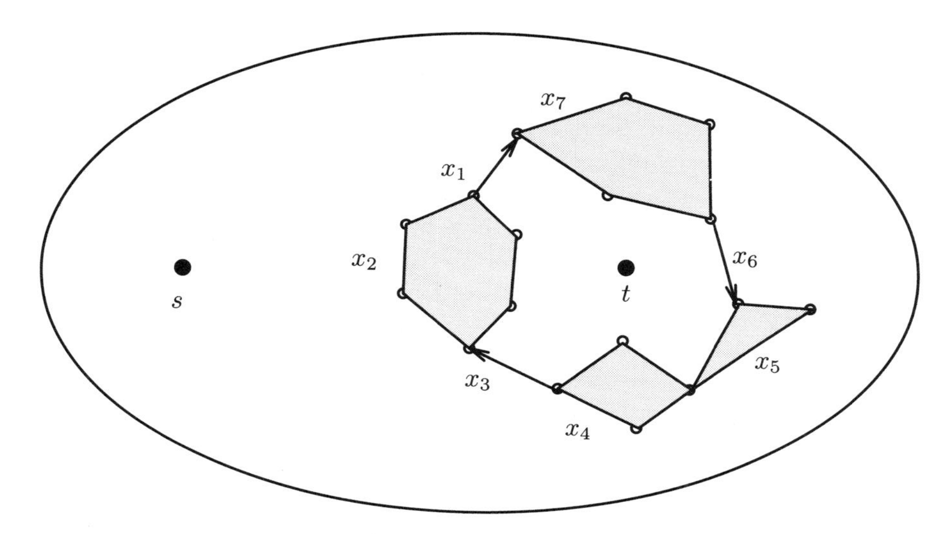

$C = [x_1, x_2, ..., x_7]$

x_2, x_4, x_5, x_7 are faces in the dual graph

x_1, x_3, x_6 are edges in the dual graph

FIGURE 2. Cycle in the Dual Graph

6.1. Computing the Min-cut. When the graph has vertex as well as edge capacities, a cut is not just a set of edges, but a subset $S \subseteq E \cup V$ with the property that every path from s to t contains an element of S. A minimum cut is defined to be a set S of minimum capacity. In the dual graph, a cut corresponds to a set of edges and faces (that correspond to the vertices in the cut). These edges and faces can be "linked" together (see Fig. 2) and induce a "linked" cycle in the dual graph that separates the faces corresponding to s and t.

In the dual graph we define a new shortest path computation as follows:

DEFINITION 6.1.1. *We are given a planar dual graph $\mathcal{D}$ with a cost $c(e_i)$ on each edge e_i, and a cost of $c(f_j)$ on each face f_j (this cost is the capacity of the corresponding primal vertex). We define a linked cycle to be a sequence of edges and faces $[x_1; x_2; \ldots; x_k]$ so that each x_i and x_{i+1} share a common vertex. (See Fig. 2 for an example.) The length of a linked cycle is the sum of the costs of the edges and the costs of the faces the cycle "jumps" over (to move from one edge to another). The shortest linked cycle is defined to be the linked cycle with the least length.*

Under this definition, the minimum cut corresponds to the shortest "linked" cycle in the dual graph that separates s from t. In [**34**] we show how to modify the dual graph so that such a cycle can be computed efficiently (the key point is that we wish to maintain planarity, to permit the use of Frederickson's shortest path algorithms).

We have now reduced the problem of finding the minimum cut in a planar graph with vertex capacities to that of finding the minimum length cycle separating s from t in a new planar graph, $\mathcal{D}'$, that has only edge capacities. The efficiency of computing this cycle varies with respect to whether the source and sink are on the same face, or whether the graph is directed.

By an application of the above idea, we can show that for st-graphs we can obtain a fast algorithm for finding the min-cut. The bounds for the parallel algorithm follow from the proof of Theorem 4.1.2.

THEOREM 6.1.2. *We can compute the* value *of the max-flow in a st-graph (directed or undirected) in $O(n\sqrt{\log n})$ time. Moreover, we can implement this algorithm in $O(\log^3 n)$ time using $O(n^{1.5})$ processors on an EREW PRAM.*

In [**49**] it was shown that the minimum cut (or the value of the max flow) can be computed efficiently even when the vertices s and t are not on the same face in an undirected planar graph. Using Frederickson's algorithm for shortest paths in planar graphs as a subroutine, one can obtain a running time of $O(n \log n)$. We note that by using the "jumping" over faces idea we get an $O(n \log n)$ time algorithm for computing the minimum cut in the graph even when the vertices have capacities.

The problem of finding a minimum cut (between a single source-sink pair) in the directed graph case is considerably harder than in the undirected case and was dealt with by [**29**]. Recently, an elegant technique was developed by [**41**] to find the minimum cut (this simplifies the procedure of [**29**]). In [**34**] it is shown that an appropriate modification of the method is able to find the minimum cut even when vertex capacities are present. (This is because vertex capacities cause an altered structure to the min-cut.)

THEOREM 6.1.3. *The minimum cut in a directed planar graph can be found in $O(n^{1.5} \log n)$. A parallel implementation uses $O(n^{1.5})$ processors and $O(\log^4 n)$ time on the EREW PRAM.*

The parallel complexity follows from the proof of Theorem 4.1.2. The additional $O(\log n)$ factor is due to the binary search for the value of the min-cut.

6.2. Computing the flow function. The main difficulty in computing the flow function with vertex capacities is that the potential function computed in the dual graph with "jumping over faces" is not *consistent.* As a consequence, computing the flow function becomes much more complicated than in the case where there are only edge capacities.

The first case we deal with are st-graph's (both undirected and directed). In [**34**] an $O(n \log n)$ implementation of the "uppermost path" algorithm due to Ford and Fulkerson [**7**] is given (that handles vertex capacities as well).

We give an $O(n \log n)$ algorithm to compute a valid flow function in an undirected st-graph that has vertex capacities. The algorithm can be extended to the case of directed st-graphs quite easily by using the ideas in [**26**] to find the directed uppermost path in each iteration. For details on the uppermost path algorithm we refer the reader to [**26**] and [**7**] (see also [**43**]).

We will briefly outline the modifications to the algorithm to handle vertex capacities. The algorithm begins by pushing flow through the uppermost path from s to t (see Fig. 3).

The capacity of the uppermost path is defined to be the least residual capacity of either an edge or a vertex. At least one edge or vertex on the uppermost path gets saturated by pushing a flow of value equal to the capacity of the path. The saturated edge/vertex is deleted from the graph, and the process is repeated using the uppermost path in the residual graph until s is disconnected from t.

Care needs to be taken to make the uppermost path *simple* each time we delete the saturated edge or vertex. The reason for this is the presence of vertex capacities in the graph. In the case of only edge capacities, pushing a flow of value equal to the capacity of the uppermost path does not violate any capacity constraint. Suppose there are vertex capacities present and the path is non-simple at a vertex that has a capacity. Pushing a flow of value f on this path actually increases the incoming flow to this vertex by at least $2f$ units, which could cause a violation in the capacity constraint of this vertex. This is the main modification to the algorithm presented in [**26**]. The algorithm discards pieces of the graph in making the path simple at each pushing step. By the following proposition we can see that the value of the flow function computed by this modified uppermost path algorithm is the same as the value of the min-cut.

PROPOSITION 6.2.1. *The process of making the augmenting path simple at each step does not decrease the amount of flow pushed on that augmenting path.*

THEOREM 6.2.2. *A maximum flow function can be computed in $O(n \log n)$ time for the case of st- graphs, even when the vertices have capacities.*

A parallel algorithm to find the max-flow in an st-graph (directed and undirected) that works by canceling the spurious cycles in the graph is also given in

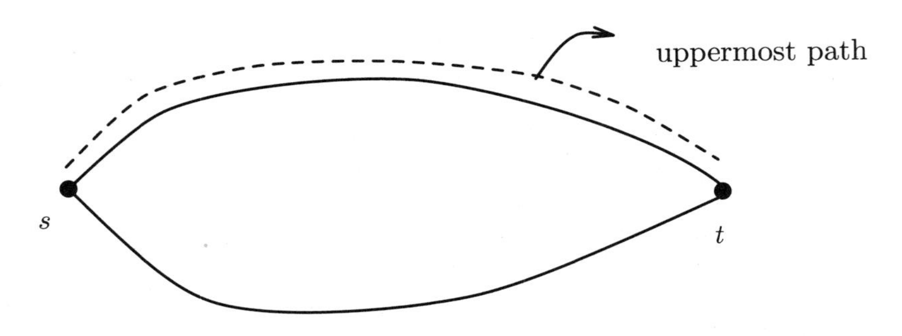

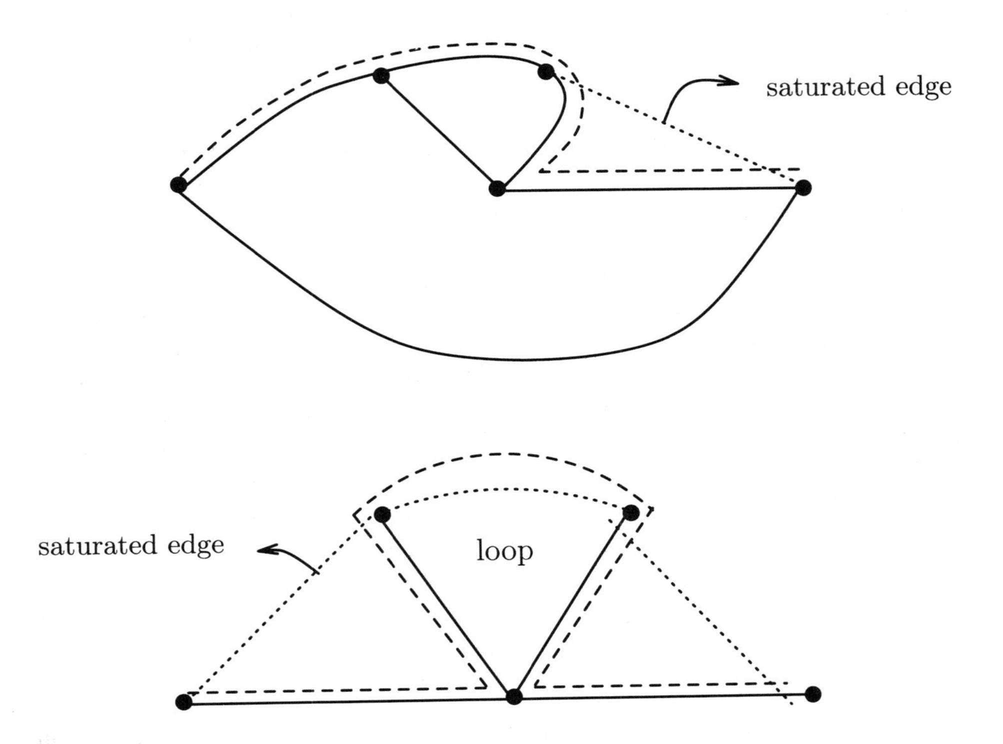

FIGURE 3. Uppermost path may be non-simple

[**34**]. (This is described in the next subsection.) A sequential implementation of the parallel algorithm takes $O(n\sqrt{\log n})$ time without counting the time for the step that requires sorting. (Thus we could obtain an $O(n\sqrt{\log n})$ time randomized algorithm by using the fast randomized sorting algorithm due to [**9**] that runs in $O(n\sqrt{\log n})$ expected time.)

If the source and sink are not on the same face, then we first find the value of the max-flow by the parametric search technique. The problem then reduces to a fixed demand problem. If there are many sources and sinks in the graph (with fixed demands), then we reduce it to the problem of computing a circulation. This is done similarly to [**41**] by "piping" the flow back from the sinks to the sources, via a path that must not go through any capacitated vertices.

An efficient algorithm for computing a circulation when edges have lower bounds and vertices are capacitated is also given in [**34**]. This tackles the case when the demand of each source and sink is known. We show how to compute a circulation that will satisfy the demands, or determine that a feasible circulation does not exist.

In the case of vertex capacities, defining the flow through an edge to be the difference of the potentials of the faces on each side as computed in $\mathcal{D}'$ (with jumping over faces), yields a flow function that may violate vertex capacity constraints.

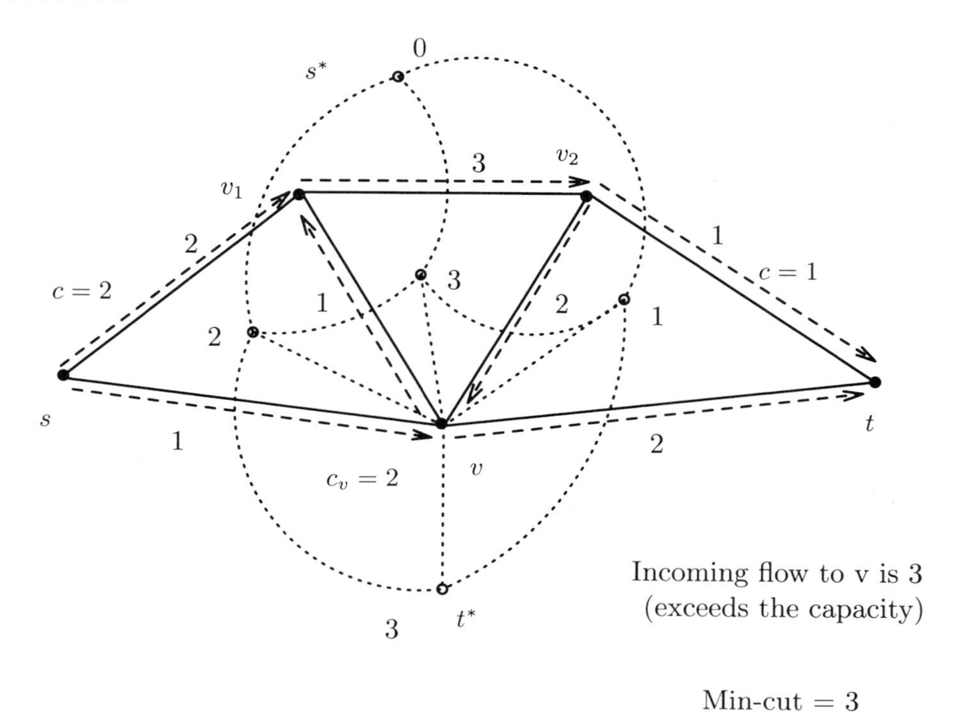

FIGURE 4. Example to show violation of vertex capacities

6.3. The parallel algorithm for st-graphs. We develop a two phase parallel algorithm to find a valid max-flow in the case of st-graphs. We only give an informal overview of the algorithm. In the first phase, we compute the potential of each face by a shortest path computation in the dual graph with s^* as the source. This is done with jumping over faces permitted (which can be reduced to a shortest path computation, as was shown in Section 3). If there are no capacitated vertices then clearly this yields a valid flow function. In certain cases, it may also happen that this procedure yields a valid flow function even in the presence of vertex capacities. In general, it does not yield a valid flow function (as the earlier example showed) due to the presence of "spurious cycles".

In the second phase we show how to fix all the "unhappy vertices" (that have excess flow through them). To motivate the second phase let us see what goes "wrong" when we compute potentials via jumping over faces. Consider a vertex v that has capacity c_v, and its incident faces. We assume that the incident faces have potential values $p_0^v, p_1^v, \ldots, p_{d(v)-1}^v$ (where $d(v)$ is the degree of v). We can assume that p_0^v is the smallest potential value and that the ordering of the faces is anticlockwise (see Fig. 5 for an example). Number the faces such that p_i^v is the potential of face i. The edge incident on v between face i and $(i+1) \mod d(v)$ is called e_i. Since the potentials were computed with "jumping over faces" we know that

$$| p_i^v - p_j^v | \leq c_v \quad \forall i, j$$

$$| p_i^v - p_{i+1}^v | \leq c_{e_i} \quad \forall i.$$

If we traverse the faces starting from face 0 in an anticlockwise direction, whenever the potential goes up it corresponds to an edge with incoming flow. The amount of incoming flow is the same as the change in potential. Correspondingly, whenever the potential goes down, it corresponds to an edge with outgoing flow. (In Fig. 6 we illustrate a vertex v with seven edges incident on it, and the corresponding potential sequence.) Clearly, each jump in the potential, either up or down, is bounded above by $\min(c_v, c_e)$ where c_e is the capacity of the corresponding edge. As we do the traversal, the total incoming flow could easily exceed c_v.

We now show a correspondence between the uppermost path algorithm and the shortest path algorithm. This is important for understanding how the potentials can be adjusted to cancel the relevant spurious cycles. The uppermost path algorithm really corresponds to growing a shortest path tree T_D from s^*. The augmenting path at each step corresponds to the "fringe" of the faces corresponding to vertices in tree T_D at various stages of a Dijkstra shortest path computation. When the fringe is non-simple, the uppermost path is also non-simple and needs to be made simple.

The flow function computed by assigning potentials, directly corresponds to an uppermost path algorithm *without* making the path simple at each step – this is precisely what causes excess flow to go through capacitated vertices.

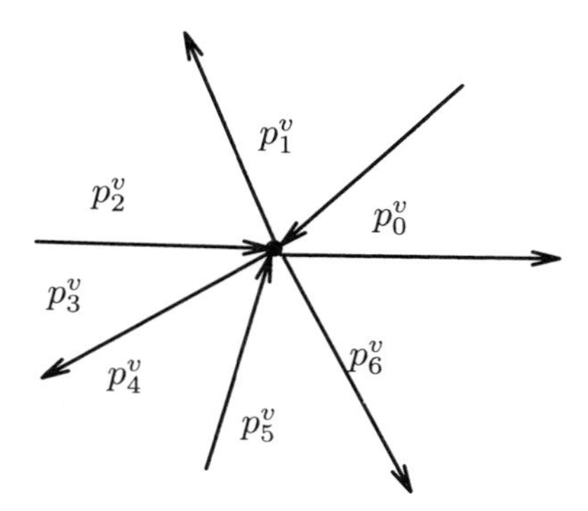

FIGURE 5. A capacitated vertex and the potentials of its incident faces

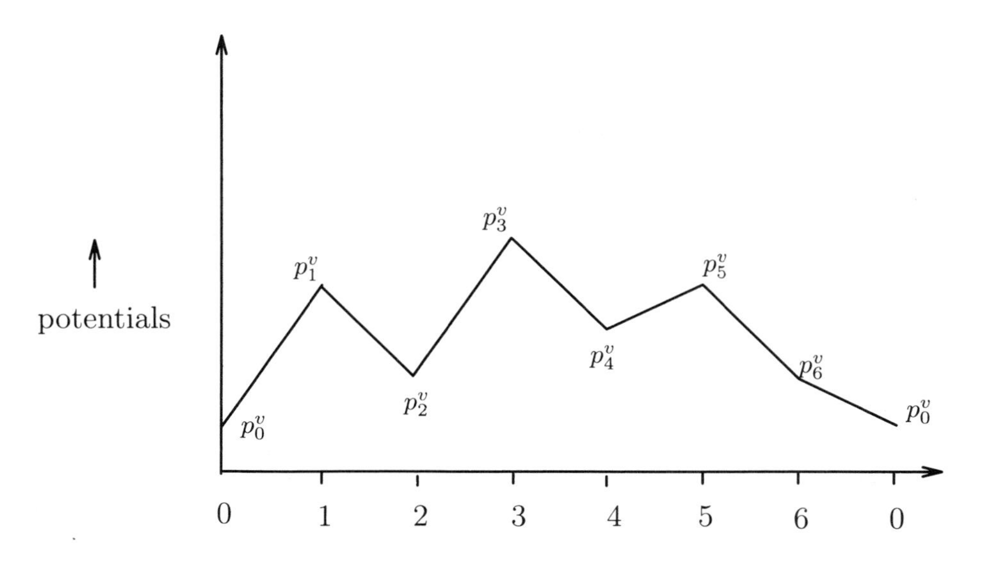

FIGURE 6. Potentials of faces incident on v

In the second phase we will try and cancel all the "spurious cycles" that cause capacitated vertices to be unhappy. The idea is to consider various snapshots of the dual tree. Examining the snapshots of the dual tree encode the various stages of an uppermost path algorithm. The non-simplicities are easy to detect and the potentials can be adjusted to cancel some spurious cycles (at least enough cycles so as to satisfy the capacity constraints of all vertices).

We now get the following theorem.

THEOREM 6.3.1. *A max-flow in an st-graph (directed and undirected) can be found in $O(\log^3 n)$ time on an EREW PRAM using $O(n^2)$ processors.*

6.4. Reduction from flow to circulations. A flow problem is transformed to a circulation problem with lower bounds on the edges. This will be done by adding new edges that will return the flow from the sinks back to the sources (as in Section 4.1). These edges will have lower bounds so as to ensure that the demands of the sources and sinks are satisfied. This reduction works for both undirected and directed graphs, but generates a directed graph. In the new graph we will compute a circulation and obtain a legal flow that satisfies the demands by removing the newly added edges.

6.5. Computing circulations. In this section we show how to use the planar separator theorem [**40**] to obtain a solution for the circulation problem when the graph contains edge capacities (upper and lower) as well as vertex capacities. We will assume that the graph is triangulated. This approach is similar to the algorithm developed by [**31**].

An overview of the algorithm:

Step 1. Find a separating cycle C of size $O(\sqrt{n})$. Let the *interior* and *exterior* of G be denoted by G_I and G_E.

Step 2. Recursively find a circulation in $G_I + C$ and $G_E + C$.

Step 3. Merge the circulations computed in Step 2, to obtain a circulation in G.

In [**34**] the following lemma is proven.

LEMMA 6.5.1. *There exists a feasible solution in $G_I + C$ and $G_E + C$.*

To combine the two circulations, we need to redirect flow on the edges of the cycle C. This is done by making $O(\sqrt{n})$ calls to the *st*-graph subroutine (see details in [**34**]), and this dominates the complexity of the algorithm.

We can prove the following:

THEOREM 6.5.2. *The complexity of computing a circulation in a planar graph with vertex capacities is $O(n^{1.5} \log n)$.*

7. NP-completeness results

There are several flow problems that are known to be NP-complete for general graphs. We show that these are NP-complete when we restrict our attention to planar graphs as well. Usually, it is unclear a priori if the restriction of planarity makes the problem computationally easier or not. There are at least two problems that come to mind that are NP-complete for general graphs, but solvable in polynomial time for planar graphs. The first is the problem of checking if a graph G contains a bipartite subgraph with at least K edges. The second is the problem of finding a max-cut in a weighted graph. Both these problems were shown to be solvable in polynomial time for planar graphs by Hadlock [**20**], even though for general graphs they are NP-complete [**10**].

On the other hand, many NP-complete problems, e.g., Hamiltonian path, 3-coloring, remain NP-complete even when restricted to planar graphs.

Consider the following flow problems. In [**2**] it was shown that the first two problems remain NP-complete even if we restrict our attention to planar graphs.

(i) **Min-edge cost flow:** Given a directed graph $G(V, E)$ and specified sets of vertices $S = \{s_1, \dots, s_k\}$ and $T = \{t_1, \dots, t_\ell\}$, capacity $c(e) \geq 0$, and price $p(e)$ for each $e \in E$, demands $R_1, \dots, R_\ell$ (for each sink) and bound B. Is there a flow function f, such that
- For each $e \in E$, $f(e) \leq c(e)$.
- For each vertex $v \in V - (S \cup T)$, the incoming flow is equal to the outgoing flow.
- The net flow into t_i is at least R_i.
- If $E' = \{e \mid f(e) \neq 0\}$, then $\sum_{e \in E'} p(e) \leq B$.

(ii) **Integral flow with multipliers:** Given a directed graph $G(V, E)$, specified sets of vertices $S = \{s_1, \dots, s_k\}$ and $T = \{t_1, \dots, t_\ell\}$, multiplier $h(v) \in Z^+$ for each $v \in V - \{S, T\}$, capacity $c(e)$ for each $e \in E$, demand R_i for sink t_i. Is there an integral flow function f, such that
- For each $e \in E$, $f(e) \leq c(e)$.
- For each $v \in V - \{S, T\}$, $\sum_{(u,v) \in E} h(v) f((u, v)) = \sum_{(v,u) \in E} f((v, u))$.
- The net flow into sink t_i is at least R_i.

(iii) **A circulation in an undirected graph:** Given an undirected graph $G(V, E)$ where each edge $e \in E$ has both a lower bound $l(e)$ and an upper bound $u(e)$ on its capacity, compute a feasible circulation f, i.e., a circulation in which
- for every edge e, $l(e) \leq f(e) \leq u(e)$.
- flow conservation is maintained in each vertex.

In other words, can the edges be oriented such that a feasible circulation exists? This problem was shown by Itai [**24**] to be NP-complete. Following the results in [**41**] (see Section 4.1), this problem can be recast in planar graphs as follows. Consider an undirected planar graph, where each edge is associated with two numbers $l(e)$ and $u(e)$. Replace each edge e by two directed edges, oppositely oriented, with weights $-l(e)$ and $u(e)$, such that there are no negative weight cycles in the graph. Suppose that $l(e) = 1$, $u(e) = k - 1$ and the graph is bridgeless. Tutte

[**53**] observed that a feasible circulation in this case exists if and only if the dual graph is k-colorable. (Using our terminology, this follows by taking the potentials on the faces modulo k.) Tutte used this correspondence between circulations and colorings to obtain an equivalent formulation of the four-color problem; he called circulations of this form in undirected graphs *nowhere-zero flow*. Jaeger [**27, 28**] showed that every bridgeless graph, (not necessarily planar), has a nowhere-zero flow where $k = 8$. This was subsequently shown by Seymour [**52**] to hold for the case $k = 6$ as well and it is open whether it holds for the case $k = 5$. (Note that it is NP-complete to decide if a planar graph has a nowhere-zero flow where $k = 4$.)

8. Conclusions and open problems

There are still many interesting open questions that are related to flow in planar graphs. Here we mention a few of them.

Computing a maximum flow with many sources and sinks.

We have presented efficient algorithms for the case where the sources and sinks belong to a small number of faces, and for the case where the demands and supplies are given as part of the input. However, this problem, in its most general setting where the sources and sinks belong to an arbitrary number of faces, is still open. We would like to take advantage of the planarity of the graph to design more efficient algorithms, sequential as well as parallel, in the case of multiple sources and sinks. In the parallel context, maximum flow in a general network was shown to be P-complete [**15**] and hence, it is widely believed not to have an efficient parallel algorithm. On the other hand, maximum flow can be reduced to maximum matching and this reduction implies an RNC algorithm when the edge capacities are represented in unary [**32, 42**]. Since even in the unary capacity case we do not have a deterministic parallel algorithm, this emphasizes the importance of solving the problem in the case of a planar network. Notice that for the restricted case of a single source, single sink, there exist NC algorithms [**22, 29**].

Computing a maximum matching in NC in a planar graph.

We mention this problem here since matchings and flows are closely related problems. The situation with computing a perfect matching in planar graphs is very intriguing. Kasteleyn [**33**] had already shown how to *count* the number of perfect matchings in a planar graph, a problem that is # P-complete in general graphs, and his methods can be implemented in NC (see e.g., [**38, 54**]) as well. Yet computing a perfect matching in NC in a planar graph remains an open problem. This situation is interesting as it contradicts the current view of the computational difficulty of counting the number of solutions versus finding a solution in combinatorial problems. It follows from the results in Section 4.1 that a perfect matching can be computed in a planar bipartite graph (see also [**41**]).

Computing a minimum cost circulation in a planar graph.

The minimum cost circulation problem is that of obtaining a circulation of minimum cost in a network whose edges have both capacities and costs per unit of flow. The problem is equivalent to the transshipment problem and has wide applicability to a variety of optimization problems.

The current best algorithm for computing a minimum cost circulation in a general graph is Orlin's algorithm [**44**]. His algorithm implies an $O(n^2 \log^{1.5} n)$ algorithm for minimum cost circulation in planar graphs. (Frederickson's shortest path algorithm [**8**] is used). A natural open question is whether the special properties of planar graphs can be exploited to obtain a faster sequential algorithm, or an NC parallel algorithm. (Notice that minimum cost circulation is P-complete since maximum flow is a special case of it).

The only algorithm that exclusively deals with the minimum cost circulation in planar graphs is by Imai and Iwano [**25**] who suggested both weakly polynomial algorithms and parallel algorithms (not NC) based on interior point methods and planar separators. Let γ denote the maximum absolute value of the cost and capacity in the graph. The running time of the sequential algorithm is $O(n^{1.594}\sqrt{\log n}\log(n\gamma))$ and the parallel time is $O(\sqrt{n}\log^3 n\log(n\gamma))$ using $O(n^{1.094})$ processors.

For planar graphs, the cost function can be interpreted as a modular function on the lattice where the cost of a face is defined to be the sum of the costs of its edges traversed in the clockwise direction. Minimizing a modular function over a lattice is a well known problem in operations research and can be solved as follows: compute the minimum cut (max flow) in a network obtained from the directed acyclic graph in which there is a 1-1 correspondence between lattice elements and closed subsets. However, this approach does not directly yield a new polynomial-time min-cost circulation algorithm, since the directed acyclic graph is too large.

We conclude by noting that the minimum cost circulation problem in planar graphs can be cast as a linear program where the variables are the potentials on the face.

The case of vertex capacities

We have shown a simple reduction for computing the minimum cut in a graph with capacitated vertices to a graph that has only edge capacities. However, this reduction holds only if there is one source and sink. If there are many sources and sinks, then it is not true that the minimum cut is equal to a collection of cycles of minimum capacity that separates the sources from the sinks in $\mathcal{D}'$, i.e., with "jumping over faces". The reason is that two cycles in this collection are not necessarily "independent" (if they share a common capacitated vertex). We conjecture that if there are many sources and sinks, then a simple reduction of the above form does not exist.

It seems that the major difficulty with vertex capacities is in computing the flow function. Suppose that we want to compute the flow function via a potential function in a similar way to [**41**]. As already pointed out, even if we use "jumping over faces" for computing the potential function, we do not necessarily get a legal circulation. (See Fig. 4). To obtain a legal circulation, a set of spurious cycles has to be identified and canceled. Can these cycles be efficiently identified? If the graph contains only one source and sink, then the spurious cycles have a more special structure. In every spurious cycle, the flow on an edge needs only to be decreased and never increased. Can the spurious cycles in this case be efficiently identified ? In the case of undirected graphs with a single source and sink, our algorithm is slower than that of [**22**]. We conjecture that the special structure of the spurious cycles will enable them to be canceled easily.

In the case of *st* graphs, the cycles have a special structure that is exploited by the parallel algorithm. We conjecture that a deterministic $O(n\sqrt{\log n})$ algorithm exists to compute the flow function for *st*-graphs that works by canceling these spurious cycles.

Another natural open problem is how to compute the flow function in parallel. We can do that only for *st*-graphs. Can that be done for more general classes of planar graphs? How difficult is it to compute a circulation in parallel (with vertex capacities)?

References

1. C. Berge and A. Ghouila-Houri, *Programming, games and transportation networks*, John Wiley, New York (1965).
2. D. Boneh, S. Khuller and J. Naor, *Flow problems that are NP-complete for planar graphs*, manuscript.
3. Y. Birk and J. B. Lotspiech, *A fast algorithm for connecting grid points to the boundary with nonintersecting straight lines*, Proceedings of the 2nd Symposium on Discrete Algorithms, pp. 465-474 (1991). (To appear in Journal of Algorithms.)
4. J. A. Bondy and U. S. R. Murty, *Graph theory with applications*, North-Holland (1977).
5. T. H. Cormen, C. E. Leiserson and R. L. Rivest, *Introduction to algorithms*, The MIT Press (1990).
6. B. Codenetti and R. Tamassia, *A network flow approach to reconfiguration of VLSI arrays*, IEEE Transactions on Computers, Vol. 40, No. 1, pp. 118-121 (1991).
7. L. R. Ford and D. R. Fulkerson, *Maximal flow through a network*, Canadian Journal of Mathematics, Vol. 8, pp. 399-404 (1956).
8. G. N. Frederickson, *Fast algorithms for shortest paths in planar graphs, with applications*, SIAM Journal on Computing, Vol.16, pp. 1004-1022 (1987).
9. M. L. Fredman and D. E. Willard, *Blasting through the information theoretic barrier with fusion trees*, Proceedings of 22nd Annual Symposium on Theory of Computing, pp. 1-7

(1990).

10. M. R. Garey and D. S. Johnson, *Computers and Intractability: A guide to the theory of NP-completeness*, Freeman, San Francisco, (1978).
11. H. Gazit and G. L. Miller, *A parallel algorithm for finding a separator in planar graphs*, 28th Annual Symposium on Foundations of Computer Science, pp. 238-248 (1987).
12. H. Gazit and G. L. Miller, *A deterministic parallel algorithm for finding a separator in planar graphs*, submitted for publication.
13. A. V. Goldberg, E. Tardos and R. E. Tarjan, Network flow algorithms, in *Paths, Flows and VLSI-Design*, Springer Verlag, pp. 101-164, (1990).
14. M. Goldberg and T. Spencer, *Constructing a maximal independent set in parallel*, SIAM Journal on Discrete Math, Vol. 2, pp. 322-328 (1989).
15. L. Goldschlager, R. Shaw and J. Staples, *The maximum flow problem is log space complete for P*, Theoretical Computer Science, Vol. 21, pp. 105-111 (1982).
16. R. E. Gomory and T. C. Hu, *Multi-terminal network flows*, SIAM Journal on Applied Math, Vol. 9, pp. 551-570 (1961).
17. F. Granot and R. Hassin, *Multi-terminal maximum flows in node-capacitated networks*, Discrete Applied Math, Vol. 13, pp. 157-163 (1986).
18. J. W. Greene and A. El-Gamal, *Configuration of VLSI arrays in the presence of defects*, Journal of the ACM, Vol. 31, No. 4, pp. 694-717 (1984).
19. D. Gusfield and R. W. Irving, *The stable marriage problem*, The MIT Press, 1990.
20. F. O. Hadlock, *Finding a maximum cut of a planar graph in polynomial time*, SIAM Journal on Computing, Vol. 4(3) pp. 221-225 (1975).
21. R. Hassin, *Maximum flows in (s,t) planar networks*, Information Processing letters, Vol. 13, page 107 (1981).
22. R. Hassin and D. B. Johnson, *An $O(n \log^2 n)$ algorithm for maximum flow in undirected planar networks*, SIAM Journal on Computing, Vol. 14, pp. 612-624 (1985).
23. J. E. Hopcroft and R. E. Tarjan, *Efficient planarity testing*, Journal of the ACM, Vol. 21, pp. 549-568 (1974).
24. A. Itai, *Two-commodity flow*, Journal of the ACM, Vol. 25 (4), pp. 596-611, (1978).
25. H. Imai and K. Iwano, *Efficient sequential and parallel algorithms for planar minimum cost flow*, Proceedings of the International Symposium SIGAL 1990, Tokyo, Japan, August 1990, Lecture Notes in Computer Science, Springer Verlag.
26. A. Itai and Y. Shiloach, *Maximum flow in planar networks*, SIAM Journal on Computing, Vol. 8, pp. 135-150 (1979).
27. F. Jaeger, *Flows and generalized coloring theorems in graphs*, Journal of Combinatorial Theory, Series B, Vol. 26, pp. 205-216, (1979).
28. F. Jaeger, *On nowhere-zero flows in multigraphs*, Proceedings, Fifth British Combinatorial Conference, Aberdeen, 1975; Congressus Numerantium XV, Utilitas Mathematica Winnipeg, pp. 373-378.
29. D. B. Johnson, *Parallel algorithms for minimum cuts and maximum flows in planar networks*, Journal of the ACM, Vol. 34, (4), pp. 950-967 (1987).
30. L. Janiga and V. Koubek, *A note on finding cuts in directed planar networks by parallel computation*, Information Processing Letters, Vol. 21, pp. 75-78 (1985).
31. D. B. Johnson and S. Venkatesan, *Using divide and conquer to find flows in directed planar networks in $O(n^{1.5} \log n)$ time*, Proceedings of the 20th Annual Allerton Conference on Communication, Control and Computing, University of Illinois, Urbana-Champaign, Ill., pp. 898-905 (1982).
32. R. M. Karp, E. Upfal and A. Wigderson, *Constructing a maximum matching in random NC*, Combinatorica, Vol. 6 (1), pp. 35-48 (1986).
33. P. W. Kasteleyn, *Graph theory and crystal physics*, Graph theory and theoretical physics, pp. 43-110, Academic Press, New York (1967).
34. S. Khuller and J. Naor, *Flow in planar graphs with vertex capacities*, Proceedings of Integer Programming and Combinatorial Optimization Conference, pp. 367-383 (1990). (To appear in Algorithmica (Special Issue on Network Flows).)
35. S. Khuller, J. Naor and P. N. Klein, *The Lattice structure of flow in planar graphs*,

UMIACS-TR-90-142, Univ. of Maryland, (1990). (To appear in SIAM Journal on Discrete Mathematics.)
36. S. Khuller and B. Schieber, *Efficient parallel algorithms for testing k-connectivity and finding disjoint s-t paths in graphs*, SIAM Journal on Computing, Vol. 20, No. 2, pp. 352-375 (1991).
37. P. N. Klein and J. H. Reif, *An efficient parallel algorithm for planarity*, Journal of Comput. and Sys. Sciences, Vol. 37(2), pp. 190-246 (1988).
38. L. Lovasz and M. D. Plummer, *Matching Theory*, North-Holland, (1986).
39. R. J. Lipton, D. J. Rose and R. E. Tarjan, *Generalized nested dissection*, SIAM Journal on Numerical Analysis, Vol. 16, pp. 346-358 (1979).
40. G. L. Miller, *Finding small simple separators for 2-connected planar graphs*, Journal of Computer and System Sciences, 32, pp. 265-279 (1986).
41. G. L. Miller and J. Naor, *Flow in planar graphs with multiple sources and sinks*, Proceedings of the 30th Annual Symposium on Foundations of Computer Science, pp. 112-117 (1989).
42. K. Mulmuley, U. V. Vazirani and V. V. Vazirani, *Matching is as easy as matrix inversion*, Combinatorica, Vol. 7 (1), pp. 105-113 (1987).
43. T. Nishizeki and N. Chiba, *Planar Graphs: Theory and Algorithms*, Annals of Discrete Math (32), North Holland Mathematical Studies.
44. J. Orlin, *A faster strongly polynomial minimum cost flow algorithm*, Proceedings of the 20th Annual Symposium on Theory of Computer Science, pp. 377-387 (1988).
45. J. Picard and M. Queyranne, *On the structure of all minimum cuts in a network and its applications*, Mathematical Programming Study, Vol. 13, pp. 8-16, (1980).
46. V. Pan and J.H. Reif, *Fast and efficient parallel solution of sparse linear systems*, Technical report 88-19, Computer Science Dept., SUNYA, 1988.
47. V. Pan and J.H. Reif, *Fast and efficient solution of path algebra problems*, J. Computer System Sciences, Vol. 38, pp. 494-510 (1989).
48. V. Pan and J.H. Reif, *The parallel computation of minimum cost paths in graphs by stream contraction*, Information Processing Letters, Vol. 40, (2), pp. 79-83 (1991).
49. J. H. Reif, *Minimum $s-t$ cut of a planar undirected network in $O(n \log^2 n)$ time*, SIAM Journal on Computing, Vol. 12, No. 1, pp. 71-81 (1983).
50. V. P. Roychowdhury and J. Bruck, *On finding non-intersecting paths in a grid and its application in reconfiguration of VLSI/WSI arrays*, Proceedings of the 1st Symposium on Discrete Algorithms, pp. 454-464 (1990).
51. V. P. Roychowdhury, J. Bruck, and T. Kailath, *Efficient Algorithms for Reconfiguration in VLSI/WSI Arrays*, IEEE Trans. on Computers, C-39:4 (Special Issue on Fault Tolerant Computing), pp. 480-489 (1990).
52. P. Seymour, *Nowhere-zero 6-flows*, Journal of Combinatorial Theory, Series B, Vol. 30, pp. 130-135 (1981).
53. W. T. Tutte, *A contribution to the theory of chromatic polynomials*, Canadian J. Math., Vol. 6, pp. 80-91 (1954).
54. V. V. Vazirani, *NC algorithms for computing the number of perfect matchings in $K_{3,3}$-free graphs and related problems*, Information and Computation, Vol. 80 (2), pp. 152-164, (1989).

Department of Computer Science, University of Maryland, College Park, MD 20742.
E-mail address: samir@cs.umd.edu

Department of Computer Science, Technion, Haifa 32000, Israel.
E-mail address: naor@cs.Technion.ac.il

DIMACS Series in Discrete Mathematics
and Theoretical Computer Science
Volume **9**, 1993

Planar Graph Coloring With an Uncooperative Partner

H. A. KIERSTEAD AND W. T. TROTTER

September 16, 1992

ABSTRACT. We show that the game chromatic number of a planar graph is at most 33. More generally, the game chromatic number of a class of graphs is bounded whenever the class is closed under minors and does not contain all graphs. Our proof uses the concept of p–arrangeability, which was first introduced by Guantao and Schelp in a ramsey theoretic setting.

1. Introduction

Let $\mathbf{G} = (V, E)$ be a finite graph, and let X be a set whose elements will be referred to as *colors*. A function $c : V \to X$ is called an *proper coloring* (or just *coloring* for short) if $c(x) \neq c(y)$ whenever x and y are distinct nodes from V with $xy \in E$. If $|\{c(x) : x \in V\}| = t$, the coloring c is also called a t*–coloring*. The *chromatic number* of $\mathbf{G}$, denoted $\chi(\mathbf{G})$, is the least positive integer t for which there exists a coloring c of $\mathbf{G}$ using a set X with $|X| = t$ as the set of colors.

In this paper, we will be concerned primarily with planar graphs. Because it is important to the spirit of results which follow, we note that there is an elementary (and very fast) algorithm for coloring a planar graph with 6 colors. By Euler's formula, a planar graph always has a node of degree at most 5. Given a graph $\mathbf{G} = (V, E)$ with n nodes, we can then label the nodes $x_1, x_2, \ldots, x_n$ so that for each $i = 2, 3, \ldots, n$, there are at most 5 neighbors of x_i in the set $\{x_j : 1 \leq j < i\}$. The graph can then be 6–colored by applying First-Fit to the nodes in the order of their subscripts in this labelling, i.e., a node is colored with

1991 *Mathematics Subject Classification.* 05C35.
Key words and phrases. Graph coloring, planar graph, chromatic number, algorithm.
The research of the first author is supported in part by ONR grant N00014-85K-0494.
The research of the second author is supported in part by NSF grant 89-02481.
This is an extended abstract of a paper which has been submitted for publication elsewhere.

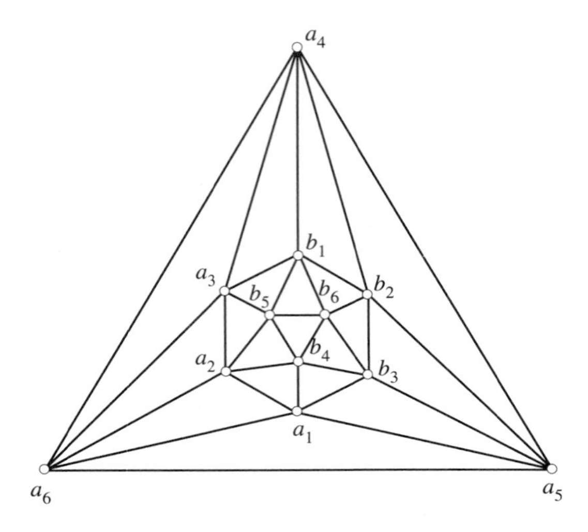

FIGURE 1.

the least positive integer distinct from the colors given to those neighbors which precede it in the labelling.

We now consider a modified graph coloring problem posed as a two-person game, with one person (Alice) trying to color a graph and the other (Bob) trying to prevent this from happening. Let $\mathbf{G} = (V, E)$ be a graph, let t be a positive integer, and let X be a set of colors with $|X| = t$. Alice and Bob compete in a two-person game lasting at most $n = |V|$ *moves*. They alternate turns with Alice having the first move. A move consists of selecting a previously uncolored node x and assigning it a color from X distinct from the colors assigned previously (by either player) to neighbors of x. If after n moves, the graph is colored, Alice is the winner. Bob wins if an impass is reached before all nodes in the graph are colored, i.e., for every uncolored node x and every color α from X, x is adjacent to a node having color α. The *game chromatic number* of a graph $\mathbf{G} = (V, E)$, denoted $\chi_g(\mathbf{G})$, is the least t for which Alice has a winning strategy. This parameter is well defined, since Alice always wins when $t = |V|$.

Example. Consider the planar graph shown in Figure 1. This graph has game chromatic number 6. To see that the game chromatic number is at least 6, here is a winning strategy for Bob if the set X of colors is $\{1, 2, 3, 4, 5\}$. Note that for each $j = 1, 2, \dots 6$, the two-element set $\{a_j, b_j\}$ is a dominating set, i.e., every other node in the graph is adjacent to at least one of these two nodes. Each time Alice colors a node from $\{a_j, b_j\}$, say with color α, Bob responds by assigning color α to the other node in this set. It follows that α cannot be used by either player to color any other node in the graph. We leave it as an exercise to show that the game chromatic number is at most 6.

The *game chromatic* number of a family $\mathcal{F}$ of graphs, denoted $\chi(\mathcal{F})$, is then defined to be $\max\{\chi_g(\mathbf{G}) : \mathbf{G} \in \mathcal{F}\}$, provided this value is finite; otherwise, we

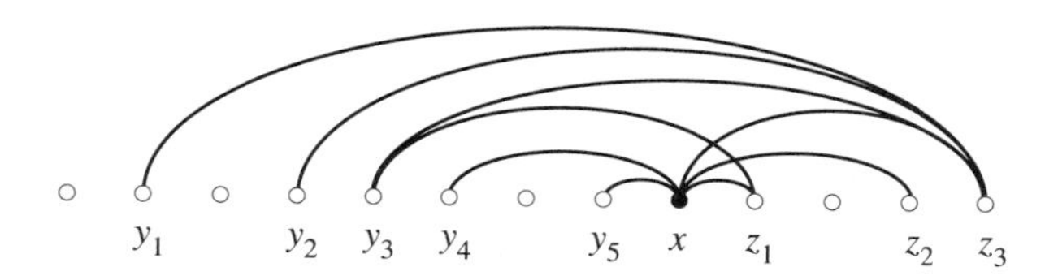

FIGURE 2.

say that $\chi_g(\mathcal{F})$ is infinite.

The concept of game chromatic number was introduced by Bodlaender [**1**] who showed that the game chromatic number of the family of trees is at least 4 and at most 5. In [**5**], Faigle, Kern, Kierstead and Trotter show that the game chromatic number of the family of trees is 4. In this paper, it is also shown that the family of bipartite graphs has infinite game chromatic number.

With these remarks as background, we can now state the principal result of this paper.

THEOREM 1.1. *The game chromatic number of the family of planar graphs is at most* 33. □

Furthermore, we will produce a very fast procedure for implementing the winning strategy. As an added bonus, we obtain results of the same form for any proper minor closed class of graphs.

2. Arrangeability and Ramsey Theory

Let $\mathbf{G} = (V, E)$ be a graph and let L be a linear order on the node set V. For each node $x \in V$, we define the *back degree of x relative to L* as $|\{y \in V : xy \in E$ and $x > y$ in $L\}|$. The *back degree* of L is then the maximum back degree of the nodes relative to L. The graph $\mathbf{G} = (V, E)$ is said to be *k–degenerate* if there is a linear order L on V which has back degree at most k. If $\mathbf{G}$ is k–degenerate, then $\chi(\mathbf{G}) \leq k + 1$, since First-Fit will use at most $k + 1$ colors when the nodes are processed in the linear order which witnesses that the graph is k–degenerate.

Again, let L be a linear order on the node set V of a graph $\mathbf{G} = (V, E)$, and let $x \in V$. We define the *arrangeability of x relative to L* as $|\{y \in V : y \leq x$ in L and there is some $z \in V$ with $yz \in E$, $xz \in E$ and $x < z$ in $L\}|$. In Figure 2, we illustrate a linear order L on the nodes of a graph. In this example, the back degree of the node x is 2. The set $S = \{x, y_1, y_2, y_3\}$ shows that the arrangeability of x relative to L is 4.

The *arrangeability* of L is then the maximum value of the arrangeability of the nodes relative to L. Following G. Chen and R. Schelp [**3**], we say that the graph

$\mathbf{G}$ is *p–arrangeable* if there is a linear order L on the nodes having arrangeability at most p.

PROPOSITION 2.1. *A p–arrangeable graph $\mathbf{G} = (V, E)$ is p–degenerate.*

PROOF. A linear order L on V which has arrangeability at most p also has back degree at most p. □

We now present a brief discussion of the ramsey theoretic problems which led Chen and Schelp [**3**] to introduce and investigate the concept of p–arrangeability. Let $\mathbf{G} = (V, E)$ be a graph. Define the *ramsey number* of $\mathbf{G}$, denoted $r(\mathbf{G})$, as the least positive integer t so that if the edges of a complete graph $\mathbf{K}_t$ on t nodes are colored with two colors, then there is always a monochromatic copy of $\mathbf{G}$. If $|V| = n$, then the ramsey number $r(\mathbf{G})$ satisfies:

$$2n - 1 \leq r(\mathbf{G}) \leq \binom{2n-2}{n-1}.$$

The lower bound in this inequality is trivial, and the upper bound is just the well known bound of Erdős and Szekeres for the ramsey number $r(\mathbf{K}_n)$. On the one hand, the exponential form of this upper bound is correct in the sense that $r(\mathbf{K}_n) \geq 2^{n/2}$. On the other hand, there are some interesting cases where the lower bound is closer to the truth. Examples include cycles and trees.

Recall that the *arboricity* of a graph $\mathbf{G} = (V, E)$ is the least t so that the edge set E can be partitioned into t forests. The following beautiful conjecture was made 17 years ago by S. Burr and P. Erdős [**2**].

CONJECTURE 2.2. *For each positive integer a, there exists a positive constant c so that if $\mathbf{G}$ is an n–node graph having arboricity at most a, then the ramsey number of $\mathbf{G}$ satisfies $r(\mathbf{G}) \leq cn$.* □

Progress in resolving this conjecture has been slow. However, in 1983, V. Chvatál, V. Rödl, E. Szemerédi and W. T. Trotter proved [**4**] a linear bound on the ramsey numbers of graphs of bounded maximum degree.

THEOREM 2.3. *For each positive integer d, there is a positive constant c so that if $\mathbf{G}$ is an n–node graph and the maximum degree of $\mathbf{G}$ is at most d, then the ramsey number of $\mathbf{G}$ satisfies $r(\mathbf{G}) \leq cn$.* □

In [**3**], G. Chen and R. Schelp develop an interesting strengthening of Theorem 2.3.

THEOREM 2.4. *For each positive integer p, there is a positive constant c so that if $\mathbf{G}$ is an n–node graph and $\mathbf{G}$ is p–arrangeable, then the ramsey number of $\mathbf{G}$ satisfies $r(\mathbf{G}) \leq cn$.* □

In order to demonstrate that their theorem applied to important examples not covered under Theorem 2.3, Chen and Schelp [**3**] then proved the following result, which is of primary importance in this paper.

THEOREM 2.5. *Every planar graph is* 761*–arrangeable.* □

Before closing this section, we make three remarks concerning Theorem 2.5 and the concept of arboricity. First, it is not immediately clear to us why planar graphs are p–arrangeable for *any* value of p, regardless of how large p is taken to be. Second, the proof of Theorem 2.3 depends heavily on Szemerédi's regularity lemma [**8**] which he first used to resolve the Erdős/Turán conjecture: Any subset of the positive integers having positive upper density contains arbitrarily long arithmetic progressions. The regularity lemma has become a much used tool in combinatorics (see [**9**] for a short proof of the lemma), but it involves constants which are just barely finite. For this reason, Chen and Schelp were not motivated to find the least value of p for which every planar graph is p–arrangeable. Third, the family of bipartite graphs used in [**5**] to show that the family of bipartite graphs has infinite game chromatic number is also a family of graphs of arboricity 2. So bounded arboricity is not enough to bound the game chromatic number.

3. Arrangeability and Graph Coloring

Let $\mathbf{G} = (V, E)$ be a graph and let L be a linear order on V. For each node $x \in V$, we say that a subset $S \subseteq V$ is *admissible for* x if: (1) $y \leq x$ in L, for every $y \in S$, and (2) there is an injection f which maps the subset $S' = \{y \in S : xy \notin E\}$ to V so that $yf(y) \in E$, $xf(y) \in E$ and $x < f(y)$ in L, for every $y \in S'$. The *admissibility of* x *relative to* L is then defined as the maximum size of a subset S which is admissible for x, and the *admissibility of* L is the maximum value of the admissibility of the nodes relative to L. A graph $\mathbf{G} = (V, E)$ is m*–admissible* if there is a linear order L on V which has admissibility at most m. For the example given in Figure 2, note that the set $S = \{x, y_2, y_3, y_4, y_5\}$ is admissible for x relative to L.

The following results are immediate.

PROPOSITION 3.1. *An* m*–admissible graph is* m*–degenerate.* □

PROPOSITION 3.2. *A* p*–arrangeable graph is* $2p$*–admissible.* □

PROPOSITION 3.3. *An* m*–admissible graph is* $m^2 - m + 1$*–arrangeable.* □

Throughout the remainder of the paper, we let $[q]$ denote the set $\{1, 2, \ldots, q\}$. The next theorem explains why the the concepts of arangeability and admissibility are important in the adversarial graph coloring environment. The proof of this result appears in the full length version of this paper.

THEOREM 3.4. *Let* $\mathbf{G} = (V, E)$ *be an* m*–admissible graph, and let* $\chi(\mathbf{G}) = r$*. Then the game chromatic number of* $\mathbf{G}$ *is at most* $rm + 1$*.* □

Recall that a graph $\mathbf{G} = (V, E)$ is *chordal* (also *triangulated* or *rigid circuit*) if it has no induced cycles of length 4 or longer. Equivalently, a graph is chordal if every induced subgraph contains a node whose neighborhood is a complete subgraph. Such a node is called a *simplicial* node. It follows that if $\mathbf{G} = (V, E)$ is a chordal graph and $|V| = n$, then the nodes of a chordal graph can be linearly ordered as $x_1, x_2, \dots, x_n$ so that for each $i \in [n]$, the neighborhood of x_i is a complete subgraph in the subgraph induced by $x_1, x_2, \dots, x_i$. Such a linear order is called a *perfect elimination scheme* (see Golumbic [**6**], for example). Hajnal and Suranyi [**7**] showed that rigid circuit graphs are perfect. As is well known, given a perfect elimination scheme, the First Fit algorithm colors the graph with ω colors, where ω denotes the maximum clique size of $\mathbf{G}$.

COROLLARY 3.5. *A chordal graph with maximum clique size ω is ω–admissible and ω–arrangeable. Therefore, it has game chromatic number at most $1 + \omega^2$.*

PROOF. Let $L = \{x_1, \dots, x_n\}$ be a perfect elimination scheme. Then the admissibility and the arrangeability of L are at most ω. This follows from the fact that if x is a node in the graph and if S is either an admissible set or an arrangeable set for x, then S induces a complete subgraph. □

4. Planar Graphs are 8–Admissible

In this section, we present the following theorem, which also yields an improvement of the bound in Theorem 2.5. Again, the proof of this result is given in the full length version of this paper.

THEOREM 4.1. *Let $\mathbf{G} = (V, E)$ be a planar graph. Then there is a linear order L of the node set V which has back degree at most 5, admissibility at most 8 and arrangeability at most 10.* □

COROLLARY 4.2. (*Theorem 1.1*) *The game chromatic number of a planar graph is at most 33.* □

COROLLARY 4.3. *If $\mathbf{G}$ is an outerplanar graph, then there exists a linear order L on the node set which has back degree at most 2, arrangeability at most 3 and admissibility at most 3.* □

COROLLARY 4.4. *The game chromatic number of an outerplanar graph is at most 10.* □

5. Lower Bounds

In this section, we produce lower bounds on the game chromatic number, admissibility, and arrangeability of planar graphs. The result for admissibility is tight, and the gap for arrangeability is modest, but we leave a relatively large gap with our lower bound on game chromatic number.

In Section 1, we gave an example of a planar graph with game chromatic number 6. We can do just a bit better.

THEOREM 5.1. *The game chromatic number of the class of planar graphs is at least 8.* □

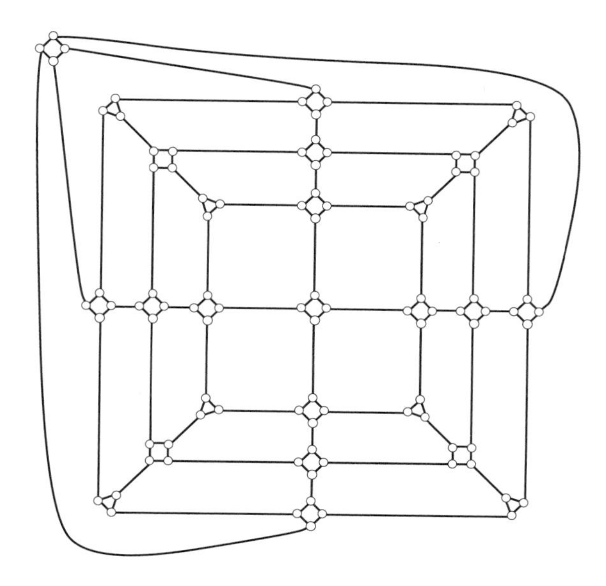

FIGURE 3.

In the argument to follow, we let $\mathbf{G}^d$ denote the *planar dual* of the graph $\mathbf{G}$.

THEOREM 5.2. *There is a planar graph with admissibility* 8.

PROOF. Consider the planar graph $\mathbf{G}$ shown in Figure 3.

Observe that $\mathbf{G}$ is constructed as follows. Begin with the 8 node cube. Add the 6 nodes and 12 edges of the planar dual of the cube. Perform a $Y - \Delta$ transformation at each node of degree 3. Insert a 4–gon at each place where edges of the cube and its dual cross. Now let $\mathbf{H} = \mathbf{G}^d$ denote its *planar dual.* For each node x in $\mathbf{H}$, let $N(x)$ denote the set of neighbors of x. It is straightforward to verify the following properties of $\mathbf{H}$.

(i) There are 24 nodes of degree 8.
(ii) Each node of degree 8 is adjacent to four other nodes of degree 8, three nodes of degree 4 and one node of degree 3.
(iii) If x is a node of degree 8 and z_1, z_2 and z_3 are its three neighbors of degree 4, then there are three nodes x_1, x_2 and x_3 so that for each $i \in [3]$:

 (a) The degree of x_i is 8;
 (b) x and x_i are not adjacent; and
 (c) x_i and z_i are adjacent.

Now let L be any linear order on the nodes of $\mathbf{H}$, and let x be the last node of degree 8 to occur in L. We show that the admissibility of x in L is at least 8. Let y_1, y_2, y_3 and y_4 be the four neighbors of x which have degree 8. Note that $y_i < x$ in L for each $i \in [4]$. Now let $\{x_1, x_2, x_3, z_1, z_2, z_3\}$ be the set of 6 nodes satisfying the third property listed above. For each $j \in [3]$, let $u_j = z_j$ if $z_j < x$ in L; otherwise, let $u_j = x_j$. Finally, let w be the unique node of degree 3 adjacent to x. Set $v = w$ if $w < x$ in L; otherwise, set $v = x$. It follows that the set $S = \{y_1, y_2, y_3, y_4, u_1, u_2, u_3, v\}$ is admissible for x. □

COROLLARY 5.3. *There is a planar graph with arrangeability at least* 8.

PROOF. Let **G** be any planar graph with admissibility 8. Form a planar graph **H** from **G** as follows. For each edge $e = xy$ in **G** add 7 new nodes of degree 2 each adjacent to both x and y. We claim that the arrangeability of **H** is at least 8. To show this, let L be any linear order of the nodes of **H**. Then let M be the restriction of L to the nodes of **G**. Choose a node x and a set S of 8 nodes from **G** so that S is admissible for x relative to M. If S is arrangeable for x relative to L, then the arrangeability of L is at least 8. Now suppose that S is not arrangeable for x relative to L. Then there is a node $y \in S$ so that xy is an edge in **G**, but none of the 8 nodes of degree 2 adjacent to both x and y added in the formation of **H** follows x in L. This implies that the back degree of x in L is at least 8, so that the arrangeability of L is at least 8. □

The reader may enjoy constructing an example to provide a lower bound on the game chromatic number of outerplanar graphs.

Exercise 5.4. There is an outerplanar graph with game chromatic number at least 5. □

6. Concluding Remarks and Open Problems

Two obvious open problems which remain are to tighten the bounds we have produced for the game chromatic number of planar and outerplanar graphs. For planar graphs, our lower and upper bounds are 8 and 33, respectively. For outerplanar graphs, our bounds are 5 and 10. A third open problem is to determine the least p for which every planar graph is p–arrangeable. We suspect that the upper bound of 10 provided in Theorem 4.1 is tight. For outerplanar graphs, the upper bound of 3 on the arrangeabilty and admissibility provided by Corollary 4.3 is tight—for both parameters.

It is easy to see that the proof of Theorem 4.1 can be modified to produce bounds on the admissibility and arrangeability of graphs belonging to any class of graphs closed under minors as long as the class does not contain all graphs. In particular, the game chromatic number of a graph is bounded in terms of its genus. We do not have a good lower bound for the inequality in Theorem 3.4. However, since this paper was completed, we have been able to show that the game chromatic number of a chordal graph is bounded by a linear function of its maximum clique size.

More generally, it seems to us to make good sense to investigate general classes of optimization problems which exhibit the key features of the uncooperative (adversarial) graph coloring problem we have studied in this paper.

References

1. H. L. Bodlaender *On the complexity of some coloring games*, Proceedings of the WG 1990 Workshop on Graph Theoretical Concepts in Computer Science, to appear.
2. S. A. Burr and P. Erdös, *On the magnitude of generalized ramsey numbers*, Infinite and Finite Sets, Vol. 1, A. Hajnal, R. Rado and V. T. Sós, eds., Colloq. Math. Soc. Janos Bolyai, North Holland, Amsterdam/London, 1975.
3. G. Chen and R. H. Schelp, *Graphs with linearly bounded ramsey numbers*, J. Comb. Theory B, to appear.
4. V. Chvatál, V. Rödl, E. Szemerédi and W. T. Trotter, *The ramsey number of a graph of bounded degree*, J. Comb. Thy. B **34** (1983), 239–243.
5. U. Faigle, W. Kern, H. A. Kierstead and W. T. Trotter, *On the game chromatic number of some classes of graphs*, Ars Combinatoria, to appear.
6. M. Golumbic, Algorithmic Graph theory and Perfect Graphs, Academic Press, New York, 1980.
7. A. Hajnal and J. Surányi, *Über die Auflosung von Graphen in vollständige Teilgraphen*, Ann. Univ. Sci. Budapest Eötvös. Sect. Math. **1** (1958), 113-121.
8. E. Szemerédi, *On sets of integers containing no k elements in arithmetic progression*, Acta Arithmetica **27** (1975), 199–245.
9. ——— *Regular partitions of graphs*, Proc. Colloque. Inter. CNRS, J.-C. Bermond, J.-C. Fournier, M. lasVergnas, and D. Sotteau, eds. (1978), 399–402.

Department of Mathematics, Arizona State University, Tempe, AZ 85287
E-mail address: kierstead@math.la.asu.edu

Bell Communications Research, 445 South Steet 2L-367, Morristown, NJ 07962, and Department of Mathematics, Arizona State University, Tempe, AZ 85287
E-mail address: wtt@bellcore.com

DIMACS Series in Discrete Mathematics
and Theoretical Computer Science
Volume **9**, 1993

Partitioning a Rectangle into Many Sub-Rectangles so that a Line can Meet only a Few

DANIEL J. KLEITMAN

September 23, 1992

ABSTRACT. We consider the questions: Suppose a rectangle is partitioned into subrectangles such that no [side-parallel] line meets more than n rectangles. How many blocks can the partition have? We obtain upper and lower bounds, and make conjectures with and without the side-parallel condition, and when blocks can be "rank-rectangles". A number of related questions are considered and conjectures made.

1. Introduction

Suppose we partition a rectangle in the plane into $f(n)$ subrectangles in such a manner that any line parallel to a side of the rectangle meets at most n of the subrectangles. How large can $f(n)$ be?

This question was raised by P. Erdős many years ago. A simple example shows, somewhat surprisingly, that the answer is at least exponential:

$$f(n) \geq 3 \cdot 2^{n-1} - 2. \tag{1}$$

R. Kannan, J. Lagarias and the present author were able to find an exponential upper bound of 3^n by an easy argument, and proposed proof of such bound as an Elementary Problem in the American Mathematics Monthly. This problem was solved by Richard Goldstein [**Goldstein87**] using an approach similar to that of the proposers.

1991 *Mathematics Subject Classification.* Primary 05C35.

Research supported by NSF Grant DMS-918403, AROSR Grant 86-0076 and NSA Grant MDA-904-92-H-3029.

This paper is in final form and no version of it will be submitted for publication elsewhere.

In this paper we consider three problems related to this question: that of reducing the gap between these bounds; of the analogous problem (also raised by P. Erdős) when lines in *any direction* must intersect at most n subrectangles; and of the problem obtained when we replace rectangles here by "rank-rectangles" (raised by Z. Füredi).

Our results are as follows:

We reduce the exponential upper bound for the original problem from 3^n to x^{2n} where x is the positive solution to the equation $x^5 + x^4 + x = 1$, so that $x^2 \leq 2.2413$.

When lines in any direction must cut at most n rectangles we find a construction which provides a lower bound of $4y^{n-1} - 3$ with $y = 3^{1/6}$. We have been unable to find any improvement in the upper bound for this problem over that for the first one, though the restriction on the partition in it is obviously much more stringent.

By a "rank-rectangle" we mean a rectangle from which subrectangles that each traverse it from any one side to an opposite side have been removed. Thus, a rank-rectangle represents the shape of a set of 1's in a $0-1$ matrix of rank 1.

We show that if a rectangle is partitioned into rank-rectangles so that any line parallel to a side meets at most n, then one can have at least $9 \cdot 5^{(n-1)/2} - 5)/4$ (or roughly $(2.2361)^n$ rank-rectangles in the partition for n odd, and at most $n\binom{2n-1}{n}$ of them (which has order $n^{1/2}4^n$) otherwise. The upper bound argument for this problem was suggested by Z. Füredi.

All these questions are easily visualized problems in elementary plane geometry, yet it is not always easy to find mathematical tools to bring to bear on them. The nature of the tools needed to resolve the remaining gaps between upper and lower bounds for these problems remains mysterious.

In the next section the original upper bound problem is considered. An approach is suggested that might be capable of resolving the problem completely. The argument used to obtain the bounds announced above is presented.

In the succeeding section a construction for the lines in any direction case is provided, along with a discussion of some generalizations.

We then present arguments leading to upper and lower bounds for the rank-rectangle problem. The upper bound argument makes use of a famous theorem of Bollobás, which we shall describe.

Finally we summarize some of the open questions in this area.

2. The Original Problem

We begin with a brief description of the construction which supplies our lower bound, which we conjecture to be the exact answer.

When $n = 1$, we can, of course, have only the trivial partition into one block. For $n = 2$, we can split that block into 4 rectangles by two, side-parallel, perpendicular lines, and it is easy to see that $f(2) = 4$. For $n = 3$ we can

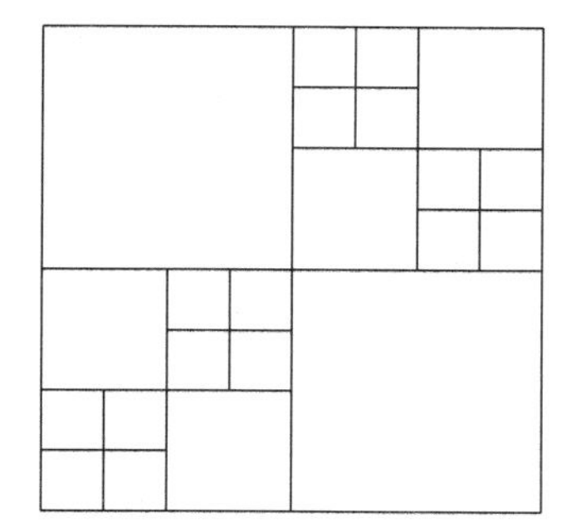

FIGURE 1. An $n = 4$ construction

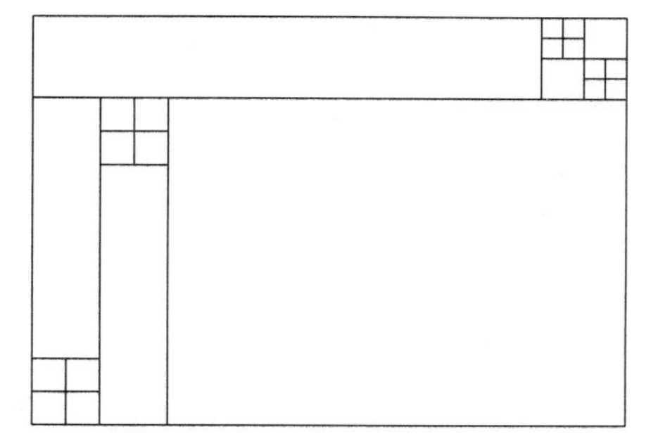

FIGURE 2. Distorted version of the previous construction

similarly split two diagonally opposite rectangles among the four, each into four rectangles, giving us $f(3) \geq 10$.

In general, we can obtain a construction for n by splitting two diagonally opposite rectangles within each of the rectangles split in forming the $n - 1$ construction, and this achieves the bound given in Equation 1. See Figure 1.

At each stage, of the four new rectangles produced in each old one that is subdivided at that stage, two remain empty and the other two continue to be subdivided at later stages.

We call a line a "cut-line" of a partitioned rectangle if it passes through the interior of the rectangle but not the interior of any block of the partition. This construction has a single "cut-line" in each side-parallel direction in each rectangle that is cut. Also, every side-parallel line through the rectangle cuts precisely n rectangles.

We can modify the construction without destroying the defining property by enlarging the two largest empty rectangles in one direction correspondly squeezing the partitioned rectangles utnil both of the latter are extremely thin. In this way, we can arrange it that most lines in one direction (except those at an edge of the original rectangle) actually cut only two subrectangles of the partition, for any n. After such distortion the partition still possesses a cut-line in one direction. See Figure 2.

In this way we can arrange it that all lines parallel to a side intersect at most n subrectagles, but all but those very near the edges intersect two rectangles in one direction and three in the other, without changing the numberof blocks in our partition. There does not seem to be a way to use these last facts to improve the lower bound here.

We now turn to the problem of finding upper bounds on the number of blocks.

The fundamental method of this section consists of cutting the partititioned rectangle into subrectangles and developing recursion inequalities based on properties of the induced partitions of the subrectangles.

We conjecture that with optimal use of this approach, using appropriate parameters to describe properties of partitions of subrectangles, the lower bound of Equation 1 can be proven to be an upper bound as well. Furthermore, we conjecture that the relevant parameters are those defined in the next paragraph. We will actually prove that using only a single parameter in our recursions, the upper bound previously mentioned can be derived.

We define the following parameters for each side-direction, i, $(i = 1, 2)$ for any partition P of a rectangle R into subrectangles:

l_i: the number of subrectangles that touch the "left" side of the rectangle

m_i: the fewest number of rectangles that meet a line, L, parallel to the left side

Ml_i: the maximum number of rectangles that meet a line parallel to the left side between that side and L.

r_i: the number of subrectangles touching the "right" side of the rectangle.

Mr_i: the maximum number of rectangles cut by a side-parallel line between L and the right side of the rectangle.

$M_i = \max(Ml_i, Mr_i)$.

Our actual argument uses only parameter M:

$$M = M_1 + M_2.$$

We will express our upper bound as $g(M)$. Note that $f(n) \leq g(2n)$ since M represents the maximum of the sum of the number of subrectangles cut by a side-parallel line in each direction.

We define the same maximum over all partitions that contain a cut line to be $gc(M)$.

Given a partition P we use the notation $M(P), m_i(P), \ldots$ to denote the M, $m_i, \ldots$ parameters of P, and write $g(P)$ for the number of blocks of P.

We will make use of two lemmas:

LEMMA 2.1. *If $m_i(P) = 2$ for any i, then $g(P) \leq gc(M(P))$.*

LEMMA 2.2. *Suppose the side-parallel line Q cuts only one subrectangle, a, that touches the left side of R. Let S be the subrectangle of R to the right of Q. Then $g(S) \leq gc(M(R))$. If Q cuts no such subrectangle and only one, b, that is adjacent to one, then $g(S) \leq gc(M(R - l))$. See Figure 3.*

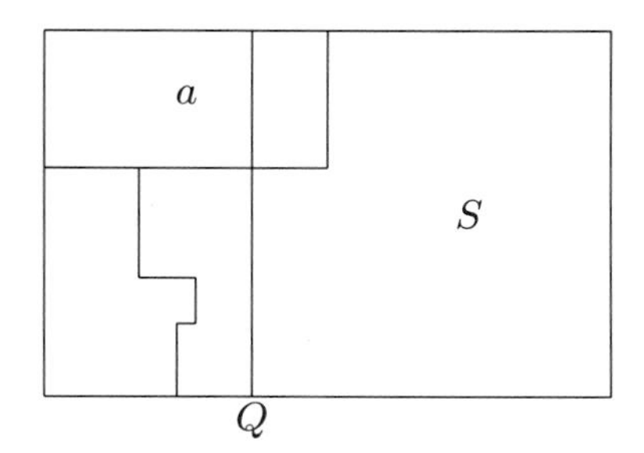

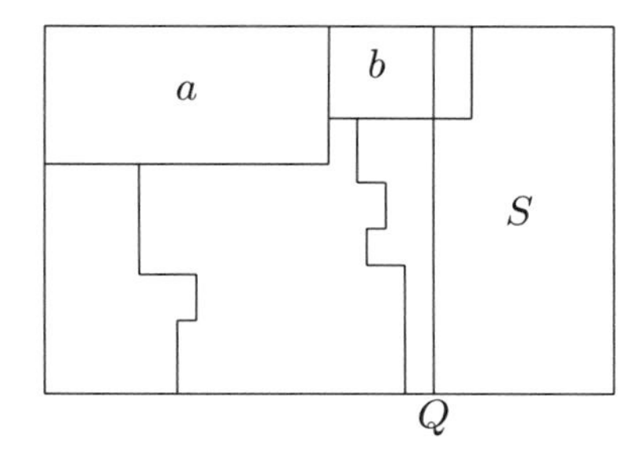

FIGURE 3. The Two Cases of Lemma 2.2

We use them to derive the following bounds:

THEOREM 2.3. $gc(M) \leq \max(2g(M-2)+2, gc(M-1)+g(M-3)+1, g(M-1)+g(M-4))$.

THEOREM 2.4. $g(M) \leq \max(gc(M-1)+gc(M-2)+1, g(M-1)+g(M-3), 2g(M-2)+g(M-6)+2)$.

THEOREM 2.5. $gc(M) \leq x^M - 1$ *with* $x^5 - x^4 - x - 1 = 0$, $x^2 = 2.2413\ldots$. *It follows that* $g(M) \leq x^{M-1} + x^{M-2} - 1$.

PROOF OF LEMMA 2.1. Suppose some line A in the 1-direction intersects only 2 rectangles, b and c, of our partition R. We shrink the region of adjacency of b and c to zero.

If b and c still exist after such shrinking, then their now common end line is a cut-line of the partition.

If one of the rectangles, say c, no longer exists after the shrinkage, then lines in the 2-direction that previously went through c now meet one less rectangle. We can therefore extend either direction-1 end-line of b across R cutting any rectangles in the way into two without increasing the number of rectangles intersecting any line in the 2-direction beyond what it was originally. This will produce a cut-line partition of R with the same parameter, unless b went all the way across R, in which case its direction-1 edge was already a cut-line. See Figure 4. □

PROOF OF LEMMA 2.2. Suppose we extend the right end of a or b completely across S creating a cut-line of S.

Then under the given conditions any line across the extension that meets new extra rectangle in S hits at least one rectangle in R that is not in S at all (and at least two such rectangles in the second case). Thus S has parameters at most M and $M-1$ in the two cases and has a cut-line. See Figure 5. □

PROOF OF THEOREM 2.3. Let the partition P or R possess a cut-line. Suppose the cut-line is in direction 1 and divides R into A and B. If $m_2(A) = 1$,

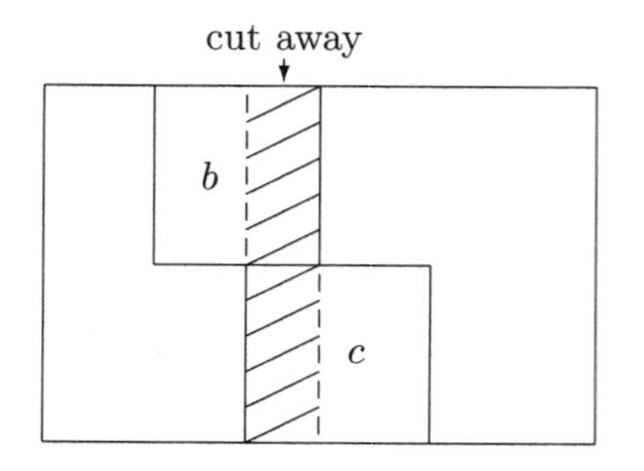

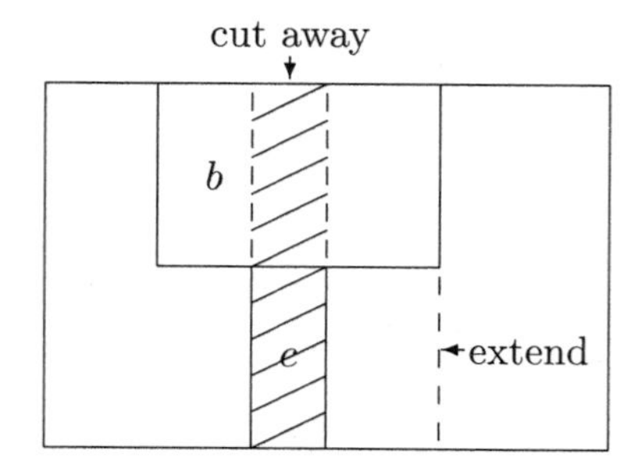

FIGURE 4. Illustrations of Proof of Lemma 2.1

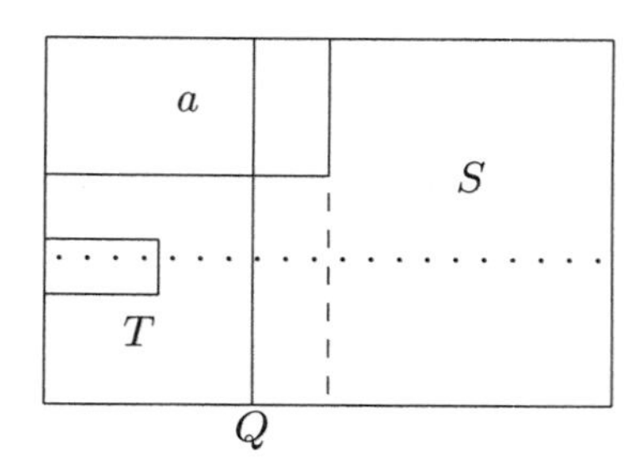

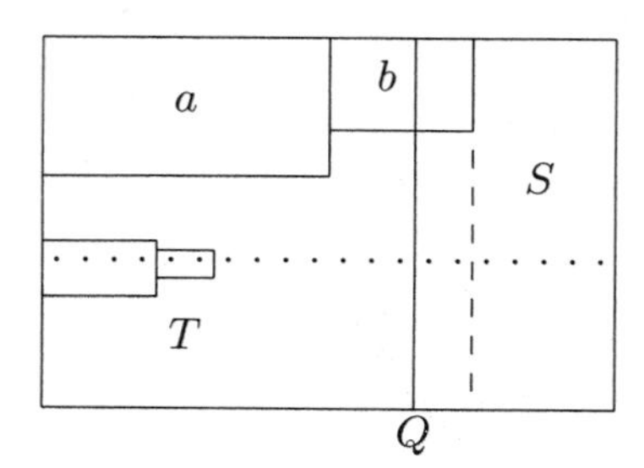

The dotted line crosses an extra subrectangle in S
but misses one [two] that is entirely in T.

FIGURE 5. Illustrations of Proof of Lemma 2.2

$m_2(B) = 1$, then removing the single rectangle that goes across A in the 2-direction leaves a region A' with $M(A') \leq M(R) - 2$, and likewise we get $M(B') \leq M(R) - 2$.

If $m_2(A) = 1$, $m_2(B) = 2$, then $M(A') \leq M(R) - 3$ and $M(B) \leq M(R) - 1$ and by Lemma 2.1, the partition of B may as well have a cut-line, giving the second alternative.

If $m_2(A) = 1$ and $m_2(B) > 2$ then we have the third alternative.

If both $m_2(A)$ and $m_2(B)$ are at least 2 then we again have the first alternative.

The three alternatives considered in this proof are illustrated in Figure 6. □

PROOF OF THEOREM 2.4. Let a be the rectangle in P touching the left side of R (which side is oriented in direction 2) that extends furthest to the right. If a's right-end-line, Q, when extended across R, cuts two rectangles of P that touch the right side of R, then by dividing R in two by line Q, we obtain $g(R) \leq g(M(R) - 1) + g(M(R) - 3)$.

Let b be the largest rectangle that touches the rightside of S and let its left-end be Q'.

Suppose Q cuts only one such rectangle, b, or the extended left end-line of b, Q', cuts only a among rectangles of P that meet the left side of R. Let

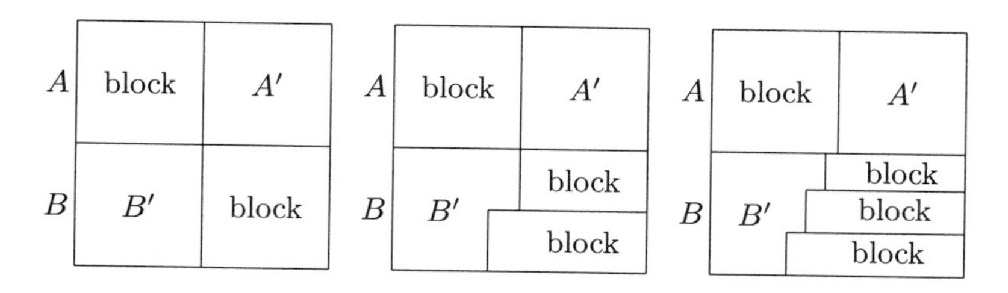

FIGURE 6. Illustration of Proof of Theorem 2.3

the regions to the left and right of Q other than rectangle a and the part of b to the right of Q, be A and B. If $m_1(B) = 2$, then $M(A) \leq M(R) - 1$, $M(B) \leq M(R) - 2$ and by Lemma 2.1 applied to B and Lemma 2.2 applied to A we get $g(R) \leq gc(M(R) - 1) + gc(M(R) - 2) + 1$.

If $m_1(B) > 2$ and a similar result holds with left and right and Q and Q' reversed, we can cut R into three pieces by Q and Q' and we get the third alternative.

If Q cuts no rectangle that touches the right end of R, then either:

the two regions that Q cuts R into (omitting rectangle a), by Lemma 2.1; or each have m_1 value 2, which gives alternative 1 in the theorem statement;

the region to the right of Q has $m_1 > 2$ in which case we get its second alternative; in the theorem statement;

or, the right-hand region has $m_1 = 2$ the left has $m_1 > 2$. We can then invoke Lemma 2.2 to get the first alternative unless two rectangles penetrate the left-of-Q region that are distance 1 from the right side. If this happens both from the left of Q and from the right of Q, there must be at least four rectangles of P that go completely across from Q and Q' and we get the third alternative by cutting R with both Q and Q'.

The cases considered in this proof are illustrated in Figure 7. □

PROOF OF THEOREM 2.5. Of the alternatives in Theorems 2.3 and 2.4, the least stringent turn out to be:

$$gc(N) \leq gc(N-1) + g(N-3) + 1 \quad \& \quad g(N) \leq gc(N-1) + gc(N-2) + 1.$$

Substituting the second in the first we get

$$gc(N) \leq gc(N-1) + gc(N-4) + gc(N-5) + 2.$$

It is easy to verify that if we substitute

$$gc(x) \leq y^{x/2} - 1 \text{ with } y = 2.2413$$

on the right hand side of these recursions (and, alternatively, all the others in Theorems 2.3 and 2.4), we may deduce this inequality inductively on x.

That this inequality holds for small values is easily verified. □

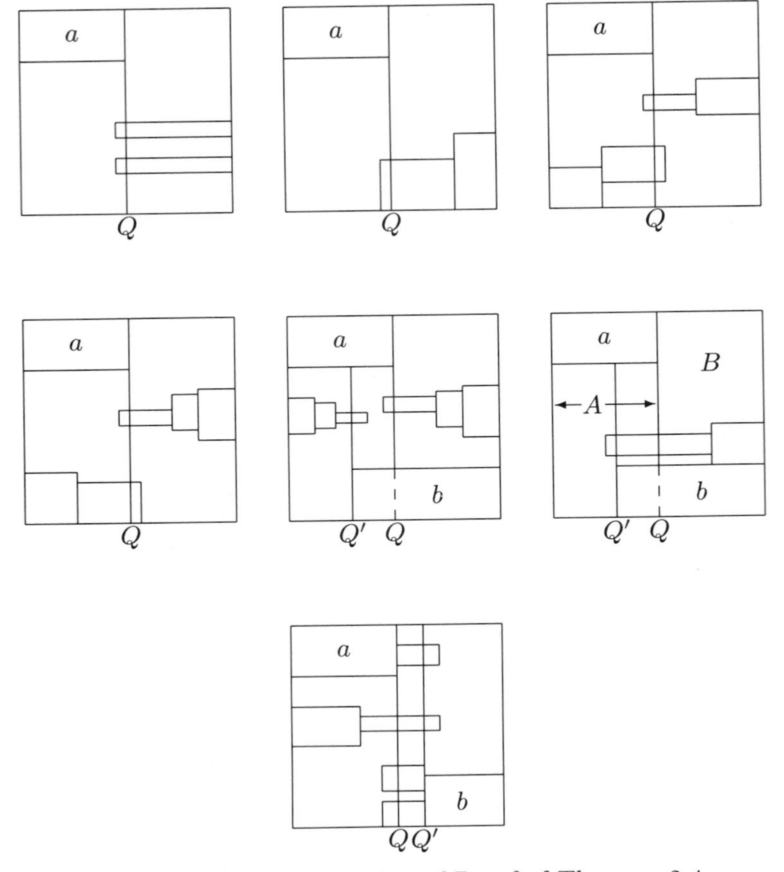

FIGURE 7. Illustration of Proof of Theorem 2.4

The upper bound obtained here is not sharp, because many of the alternatives in Theorems 2.3 and 2.4 cannot in practice be iterated. We are confronted, when we cut our rectangle into pieces, with the problem of determing behavior of partitions of Utah-shaped regions. In the given argument we use our lemma to limit their properties to those of rectangular regions with cut-lines. This representation leads to our bounds, but does not capture completely the implications of the restrictions on these regions.

3. Lines in Any Direction

If we demand of our partition that every line of every slope intersects at most n blocks, we have a stronger constraint on the partition, and expect the number of blocks to be much more limited under this restriction than under the original one.

However, we have no proof of such a statement.

The construction described in Section 2 is not suitable for the general slope problem. However, the freedom to distort a partition by squeezing some regions and expanding others along a cut-line, which had no value for the purposes of that section, can here be used to obtain partitions into subrectangles with number of blocks that is still exponential in n.

We will describe a construction that has a number of blocks given by $4 \cdot 3^{(n-1)/6} - 3$, when $n-1$ is divisible by 6. We suppose from now on that $n-1$ is divisible by 6.

Set $m = (n-1)/6$. Our construction starts with the initial "live" rectangle R with $m = 0$, and at each of m stages, divides each live rectangle into nine rectangles, three of which are live. We describe the single stage decomposition with reference to a coordinate system in which the live rectangle Z that is partitioned is a unit square. The three live rectangles produced in the partition of Z will all be much wider than high in this coordinate system, and each will be tiny in its widest dimension compared to Z.

We partition s by two vertical lines each a distance d from a side of Z, with $d < 0.001$. The thin left region is further partitioned by a horizontal line a distance d^2 above the base of Z, and the thin right region by a line a distance d^2 from the top of Z. The two d-by-d^2 regions thereby produced in diagonally opposite corners of Z are two of the three live rectangles.

We further partition the remaining large central rectangle into five rectangles, four of which fit into its corners. The four have dimensions (vertical first) of $(1/2, 1/3)$, $(1/2, 1/3 + d)$, $(1/2 - d^2, 2/3 - 2d)$, and $(1/2 + d^2, 2/3 - 3d)$. They leave a live rectangle of dimensions (d^2, d) between them.

This partition of Z into nine rectangles has the following properties:

(i) No line can pass through all three of its live rectangles.
(ii) A line can pass through all six of its dead rectangles only if it passes through at most one live one.
(iii) An approximately vertical line can intesect at most three of the dead rectangles here.

Notice that lines can pass through two live rectangles only if their slopes are approximately 1, 2/3 or 4/3.

If this construction is repeated m times, each time in the live rectangles produced at the previous stage, we find the following consequences:

First, if a line passes through only at most one live rectangle at each stage it can pass through at most six dead ones at each stage (that's all there are in any one previous live one) and one live one at the end, for a total of $6m + 1$.

Second, if at stage k, a line passes though two live rectangles, it must be essentially vertical at all successive stages. By our third property above it can pass through at most three dead rectangles within each of the two live ones at each stage after the kth, for a total of six per stage. At the mth it may pass through two live but unpartitioned rectangles, but by property 2 above it missed a dead rectangle at stage k, so that it can meet at most $6m + 1$ rectangles all together.

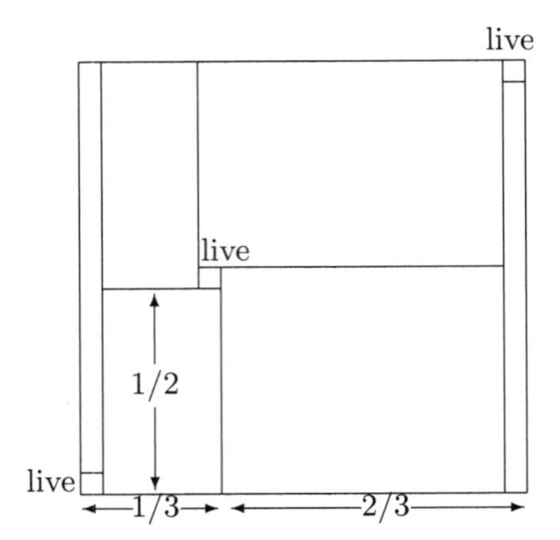

FIGURE 8. Construction stage for lines in all directions

The single stage partition is illustrated in Figure 8.

It is possible to make similar constructions with an arbitrary number of live rectangles at each stage; three seems to lead to the largest number of blocks.

It is natural to ask what conclusions hold when we relax the restriction that our lines be straight. One can ask: what sort of allowable lines destroy the exponential nature of the bound here?

It is easy to see that allowing lines that have kinks in them, consisting of k line segments of differing slopes leaves exponential dependence on n, since every such line lies in the union of k straight lines. It is perhaps interesting to ask what happens when lines obey quadratic rather than linear equations, etc.

4. Rank-rectangle Partitions

We can represent the original problem discretely by defining a matrix whose rows represent the distinct horizontal cross-sections of our partitioned rectangle, and whose columns are the distinct vertical cross-sections. By a cross-section we here mean a listing in order of the rectangles encountered by a line traversing the rectangle in the corresponding side-parallel direction.

The original problem then corresponds to determining the maximum possible number of blocks in a partition of a matrix into rectangular shaped blocks, such that there are at most n blocks in any row or column.

If $M(n)$ is a matrix minimizing the number of blocks for parameter n, then we can not only require the rows and columns to be distinct, but can insist that every row of M contains a block that occurs in no other row. Otherwise we could throw the row out of M without reducing the number of blocks.

We can then ask: What is the maximum number of rows in a matrix M that is partitioned into rectangles with the following properties?

(i) Every row and column meets at most n of the rectangles;

(ii) There is a rectangle that occurs only in the jth row for every j.

The answer is closely related to that of our original problem.

Another formulation of the same problem is as follows. Let us represent the rectangles as elements, and the rows and columns as sets. From our previous properties, each row has an element in no other row, and each row or column set has cardinality at most n. Every element in any row also lies in a column. The fact that each column set intersects each row set in a set of cardinality 1 is a consequence of the rectangular nature of the blocks, since if row i and column j contain an element of rectangle B then B must appear in the (i, j) place in M.

In the following discussion we will explore the bound on the number of rows of M and hence on the number of blocks, that follows from the conditions we have described so far:

Each row and column of M is a set of cardinality at most n. Each row has an element in no other row. Every element of a row lies in some column. Each row and column share exactly one element.

Actually these conditions are weaker than those from which we started. It was not necessary that our blocks be rectangles in order to obtain them. In fact the same conditions hold if the blocks of our original partition are what we have called "rank-rectangles".

Thus our attempt to grapple with the original problem by discretizing it and relating it to a problem on set systems led us to a more general problem: How many blocks can there be in a partition of a rectangle into rank-rectangles such that side parallel lines meet at most n of them?

Our upper bound comes from the following result of Z. Füredi, whose proof we present here:

THEOREM 4.1. *Suppose we have two collections of sets, X and Y, each member of either having at most n elements. Suppose each member of X has one element in common with each member of Y; each member of X lies in the union of the Y's, each member of X has an element in no other.*

Then X can have at most $\binom{2n-1}{n}$ members.

PROOF. For each member, A, of X, we find a member, $B(A)$, of Y, that intersects A in an element, a, unique to A among the rows.

Now let $B_j = B(A_j) - (a_j)$. Then A_j and B_j are disjoint, but A_j and B_k intersect for all j and k distinct. Each A_j has cardinality n, and each B_j has cardinality $n - 1$.

There is a theorem of Bollobás, from 1967, which tells us, that under these circumstances $M \leq \binom{2n-1}{n}$, which completes the proof. □

Bollobás' theorem has a very pretty proof, suggested by Katona, which we include for completeness.

Suppose we consider all possible orderings of all the elements in all the A_j's and B_k's. In a proportion $1/\binom{2n-1}{n}$ of these, all elements of Aj will occur before all elements of Bj. By the given conditions, these proportions cannot overlap for different j: from which the conclusion follows.

The upper bound provided by this approach is on the order of 4^n, which is worse than our previous upper bound. On the other hand, it applies to partitions into rank-rectangles. This leads to the question: can one find partitions into rank-rectangles such that side parallel lines intersect at most n of them with more blocks than one can have for partitions into rectangles?

We have not actually been able to resolve this question; however, we can beat the $3 \cdot 2^{n-1} - 2$ lower bound for partitions into rectangles.

There is another way to describe this problem. Let the *complete directed* graph on n vertices consist of edges in both directions between every distinct pair of vertices, but no loops. Let a *directed complete bipartite graph* on two disjoint vertex sets consist of all edges from one set to the other. Let the *degree*, $d(v)$, (similarly in- or out-degree) of a vertex v in a partition of edges be the number of blocks that contain any of its edges.

If for any m, we can partition the complete directed graph on m vertices into directed complete bipartite graphs having maximum in- or out-degree x, we can find a construction for this problem having on the order of $m^{(n-1)/x}$ blocks. The construction consists of using this partition for describing the dead rank-rectangles in each stage, with the diagonal elements of the matrix (which correspond to loops), providing the live rank-rectangles.

The parameters $m = 2$ and $x = 1$ give rise to our original rectangle construction.

A better construction, having $(9 \cdot 5^{(n-1)/2} - 5)/4$ rank-rectangles, for odd n, is obtained from the parameters $m = 5$, $x = 2$. We give pairs of vertex sets for five rank-rectangles in a partition of the complete directed graph on 5 vertices having $x = 2$:

$$\begin{array}{llll} A: & (1,2) & \to & (5,3) \\ B: & (2,3) & \to & (1,4) \\ C: & (3,4) & \to & (2,5) \\ D: & (4,5) & \to & (3,1) \\ E: & (5,1) & \to & (4,2) \end{array}$$

As a matrix this looks like:

$$\begin{array}{ccccc} * & B & B & D & D \\ E & * & C & C & E \\ A & A & * & D & D \\ E & B & B & * & E \\ A & A & C & C & * \end{array}$$

We have no idea whether there are parameters (m, x) which give better bounds here. This seems like an interesting combinatorial question, that may already have been explored in other contexts.

It is not necessary that a construction have regular and identical in- and out-degrees for its "live rectangle" vertices. The construction becomes more complicated and messier if in- and out- degrees are distinct and variable, but asymptotic bounds depend on the distribution of degrees as follows.

Let P be a partition of the complete directed graph on m vertices into directed complete bipartite graphs so that the degree of the ith vertex is d_i. Let z be the real root of the equation $\sum(x^{-d_i}) = 1$. Then we can obtain a construction having a number of rank-rectangle blocks which a function of n is asymptotic to that of $z^{2(n-1)}$.

Finding the best possible z value over all partitions, and all m, also seems to be an interesting problem.

5. Discussion and Open Problems

The results of the previous sections make some progress on the problems addressed, but none is completely resolved. All of them therefore remain open.

In this section we will therefore mention the open questions which seem either most amenable to being answered, or which seem most tantalizing.

The following are among these.

1. The range between upper and lower bounds we have obtained above for the three problems considered unfortunately all overlap. Thus we cannot exclude the ridiculous possibility that the solution to all have the same n dependence. It would be nice to be able to disprove this possibility.

2. Can one find a direct upper-bounding method for the lines-in-any-direction problem?

3. Are there non-Bollobasian upper bounding methods for the rank-rectangle problem, and what do they tell us?

4. What is the best construction for the rank rectangle problem?

5. Do quadratic curves rather than lines lead to exponential bounds?

6. Can one find a new approach to the original upper bound problem?

7. Can ingenuity and/or brute force solve that problem even with no new ideas?

NOTE ADDED: Z. Füredi has informed us that having heard this talk, he and E. Boros worked on these problems and were able to prove an upper bound of $c2^n$ for the original problem. They can improve this to the exact answer for $n \leq 10$.

References

[Goldstein87] Richard Goldstein, Partitioning the Unit Square, Problem E3129, Amer. Math. Monthly **94** (1987), 797–799.

DEPARTMENT OF MATHEMATICS, MASSACHUSETTS INSTITUTE OF TECHNOLOGY, CAMBRIDGE, MA 02L39

E-mail address: djk@math.mit.edu

DIMACS Series in Discrete Mathematics
and Theoretical Computer Science
Volume **9**, 1993

Disjoint Essential Circuits in Toroidal Maps

BOJAN MOHAR AND NEIL ROBERTSON

September 17, 1992

ABSTRACT. Necessary and sufficient conditions are given for a toroidal map to contain two disjoint essential cycles. The result is applied in the study of embeddings of planar graphs into general surfaces.

1. Introduction

Dirac [D] (cf. also [L]) proved that a 3-connected graph G contains no two disjoint cycles if and only if one of the following cases occurs: G is a wheel $K_1 * C_n$ ($n \geq 3$) with 3 or more spokes, $G = K_5$, or G has at least 6 vertices and contains vertices $x, y, z \in V(G)$ which cover all the edges of G. In the last case, $G = K_{3,k}$ ($k \geq 3$) or G is a graph obtained from $K_{3,k}$ by adding 1, 2, or 3 edges between the vertices in the color class of $K_{3,k}$ containing 3 vertices. Dirac's result can be generalized to arbitrary graphs. Since the removal of vertices of degree 0 or 1, and the suppression of vertices of degree 2 in a graph do not change the number of cycles, we may without loss of generality treat only the case when the minimal vertex degree is at least 3. A graph G with the minimal degree 3 or more does not contain two disjoint cycles if and only if one of the following cases occurs:

a) G has a vertex $x \in V(G)$ such that $G - x$ is a forest,
b) G has a vertex $x \in V(G)$ such that $G - x$ is a simple cycle and G has no loops at x, i.e., G is a wheel with the spokes allowed to be multiple edges,

---

1991 *Mathematics Subject Classification.* 05C10.
Key words and phrases. essential circuit, toroidal map, planar graph.
The first author was supported in part by The Ministry of Science and Technology of Slovenia and The Institute of Mathematics and Its Applications at the University of Minnesota. The second author was partially supported from NSF Grant number DMD 8903132
This paper is in final form and no version of it will be submitted for publication elsewhere

c) $G = K_5$, or
d) there are vertices $x, y, z \in V(G)$ such that $G - \{x, y, z\}$ is edgeless, there are no loops at x, y, z, and no parallel edges between $\{x, y, z\}$ and $V(G) \backslash \{x, y, z\}$. (But parallel edges between x, y, z are allowed.)

In the study of the structure of embeddings of planar graphs in the torus [MR] we bumped into the following problem: If a planar graph embedded in the torus contains no two disjoint essential circuits, when is it possible that the embedding is a closed-cell embedding? By an *essential circuit* we mean a cycle of the graph which is not contractible on the surface.

Let G be a graph embedded in a closed surface Σ. We say that G is embedded with *representativity* r, and denote this by $\rho(G) \geq r$, if every essential closed curve on the surface intersects G in at least r points. Cf. [RV] for more details on this invariant. Schrijver [S1] proved that a graph G embedded in the torus with $\rho(G) \geq r$ contains $\lfloor 3r/4 \rfloor$ pairwise disjoint essential cycles. In particular, if $\rho(G) \geq 3$, then there are two disjoint essential cycles. Although this result is best possible, it is far from a necessary and sufficient condition for a graph in the torus to contain two disjoint essential cycles. We solve this problem by characterizing graphs embedded in the torus (=*toroidal maps*) which do not admit two disjoint essential cycles (Theorems 3.1 and 3.2). A short passage then leads to the answer on our original question about the planar graphs in the torus. It should be pointed out that although the outcome is similar to the Dirac's result, the combinatorial obstructions to disjoint cycles are not of much help in the torus case.

Let us mention that the same type of questions can be posed for graphs in other surfaces. Moreover, we get a variety of problems which are of importance in the study of graph embeddings. For example, one may ask questions about the existence of disjoint essential cycles, pairwise homotopic or not (and the homotopy class fixed, or not), disjoint pairwise homologic essential cycles (homology class either fixed, or free), disjoint essential non-bounding cycles, etc. It should be pointed out that the problem of the existence of pairwise disjoint cycles of given homotopies has a "good characterization" [S2].

2. Basic definitions and some auxiliary lemmas

Graphs in this paper may have loops and multiple edges. Let G be a graph embedded in a closed surface Σ. Suppose that in Σ there is a closed curve $\gamma : S^1 \to \Sigma$ which bounds an open disk D. Let $\overline{D} := D \cup \gamma(S^1)$ be the closure of D. We say that $\overline{D}$ is a *k-patch* if $cr(\gamma, G) := |\{z \in S^1 \mid \gamma(z) \in G\}| = k$. Having a k-patch with $k = 0$, or 1, the deletion of $G \cap D$ from the graph is called a *k-reduction*. Having a 2-patch with a path in $G \cap \overline{D}$ connecting the vertices of $G \cap \partial\overline{D}$ across D, a 2-*reduction* is the operation replacing $G \cap D$ with an edge connecting the vertices of $G \cap \partial\overline{D}$ across D. Note that using 2-reductions, we can in particular eliminate all vertices of degree 2 in G (except isolated vertices with an essential loop). Having a 3-patch with $\{x, y, z\} = G \cap \partial\overline{D}$ such that in $\overline{D} \cap G$ there are paths between each pair (x, y), (x, a), and (y, z) (all of them across D), the replacement of $G \cap D$ by a new vertex w joined to x, y, z (as shown on Figure 2.1) is called a 3-*reduction*. A k-reduction ($k \leq 3$) is *non-trivial* if the graph obtained after the reduction is not isomorphic to G. In particular, the well-known ΔY-transformation is a special case of a 3-reduction if the triangle of the transformation bounds a face.

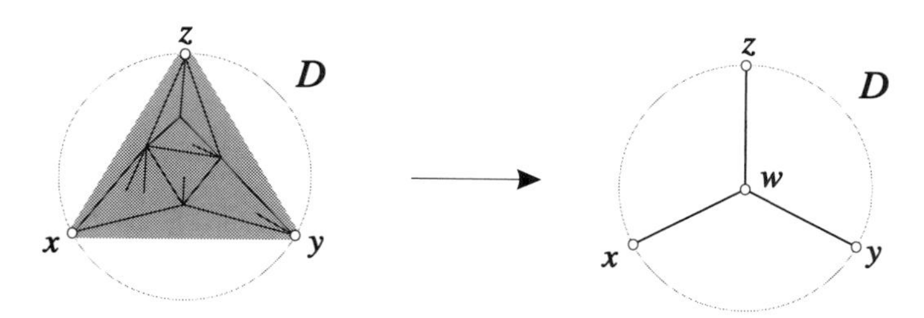

FIGURE 2.1 A 3-reduction

The following simple lemma (whose proof we omit) shows that 3-reductions preserve the maximum number of pairwise disjoint essential cycles in a graph in a surface.

LEMMA 2.1. *Let G' be obtained from G by a sequence of 0-, 1-, 2-, and 3-reductions and their inverses, and let $\mathcal{C}$ be a family of disjoint circuits of G. Then G' contains a family $\mathcal{C}'$ of disjoint circuits which are pairwise homotopic with respective circuits in $\mathcal{C}$. Moreover, the representativity of G' equals the representativity of G.*

Let K be a subgraph of G. The vertices of K of degree different from 2 (in K) are called the *main vertices* of K, and the paths in K (possibly of length 1) between main vertices, where all interior vertices are of degree 2 in K, are called *branches* of K. If B_0 is a connected component of $G - V(K)$, then the subgraph B of G consisting of B_0 and all edges between B_0 and vertices in K (together with appropriate vertices of K) is called a *relative K-component*, or a *bridge* of K. Another type of relative components (bridges) of K are the edges (together with their endpoints) which are not in $E(K)$, but connect two vertices of K. If B is a relative K-component, each vertex of $V(B) \cap V(K)$ is called a *vertex of attachment* of B to K, and each edge of B adjacent to a vertex of attachment is a *foot* of B.

If C is a cycle of G and B_1, B_2 are relative C-components, then B_1 and B_2 *overlap* (on C) if either they have three or more vertices of attachment in common, or there are four distinct vertices of attachment x_1, y_1 of B_1 and x_2, y_2 of B_2 whose order on C is x_1, x_2, y_1, y_2. If $W = x_0x_1x_2 \ldots x_k$ ($x_k = x_0$) is a closed walk in G and C its underlying subgraph of G, then a relative C-component is said to *overlap* with vertices x_i, x_j ($i < j$) on W if it has vertices of attachment

x_p, x_q such that either $p < i < q < j$ or $i < p < j < q$. We refer to [V] for a more extensive treatment of relative components.

Let G be embedded in Σ and W as above. Assume that W bounds an open disk D. If F is a face of G contained in D, then we say (with a possible slight abuse of terminology if W has some repeated vertices) that F *contains* x_i $(0 \leq i < k)$ on its boundary if F contains on its boundary either the edge $x_{i-1}x_i$, the edge $x_i x_{i+1}$ (indices modulo k), or an edge in D lying between $x_{i-1}x_i$ and $x_i x_{i+1}$ according to the local rotation at x_i.

LEMMA 2.2. *Let G be a graph embedded in a surface Σ and $W = x_0 x_1 \dots x_k$ a closed walk in G whose underlying graph C bounds an open disk D in Σ. For $i < j$ there is a face in D containing x_i and x_j on its boundary if and only if no relative C-component embedded in $D \cup C$ overlaps with x_i and x_j on W.*

Proof. If a face F in D contains x_i and x_j, then, clearly, no relative component in D overlaps with x_i, x_j. Conversely, if none of the faces $F_1, F_2, \dots, F_s$ in D, containing x_i on the boundary, also contains x_j then in the union of their boundaries there is a path from a point x_p to x_q on W. Without loss of generality $i < p < j < q$. This path is clearly contained in a relative component which is thus overlapping with x_i, x_j. □

3. Disjoint essential circuits

In this section we will state our main results whose proofs are deferred until Sections 5 and 6.

THEOREM 3.1. *Let G be a toroidal map with representativity $\rho(G) \leq 1$. Then G contains no two disjoint essential cycles if and only if the embedding of G has the structure as shown in Figures 3.1–3.2 (case $\rho(G) = 0$), or in Figures 3.3–3.5 (case $\rho(G) = 1$).*

Note: In Figures 3.3–3.5 it may be assumed that x, y, z are distinct vertices. The exact meaning of "*having the structure*" is that there is a homeomorphism of the torus on the standard "flat" torus as represented in all applicable figures (with proper standard side identifications), so that any edges of G are embedded in the shaded parts. Note that Figure 3.3 and Figure 3.4 are dual to each other and that all the others are "self-dual" structures.

FIGURE 3.1

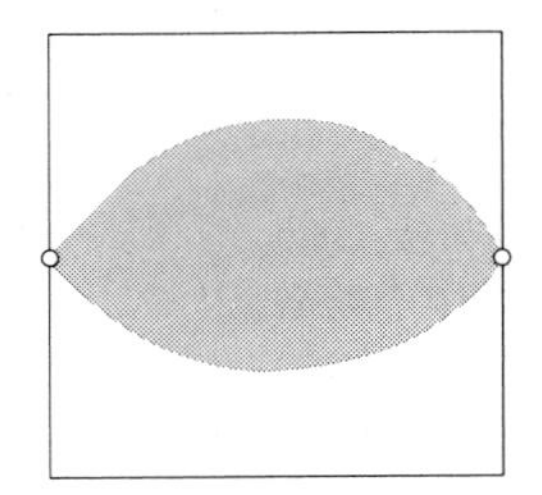

FIGURE 3.2

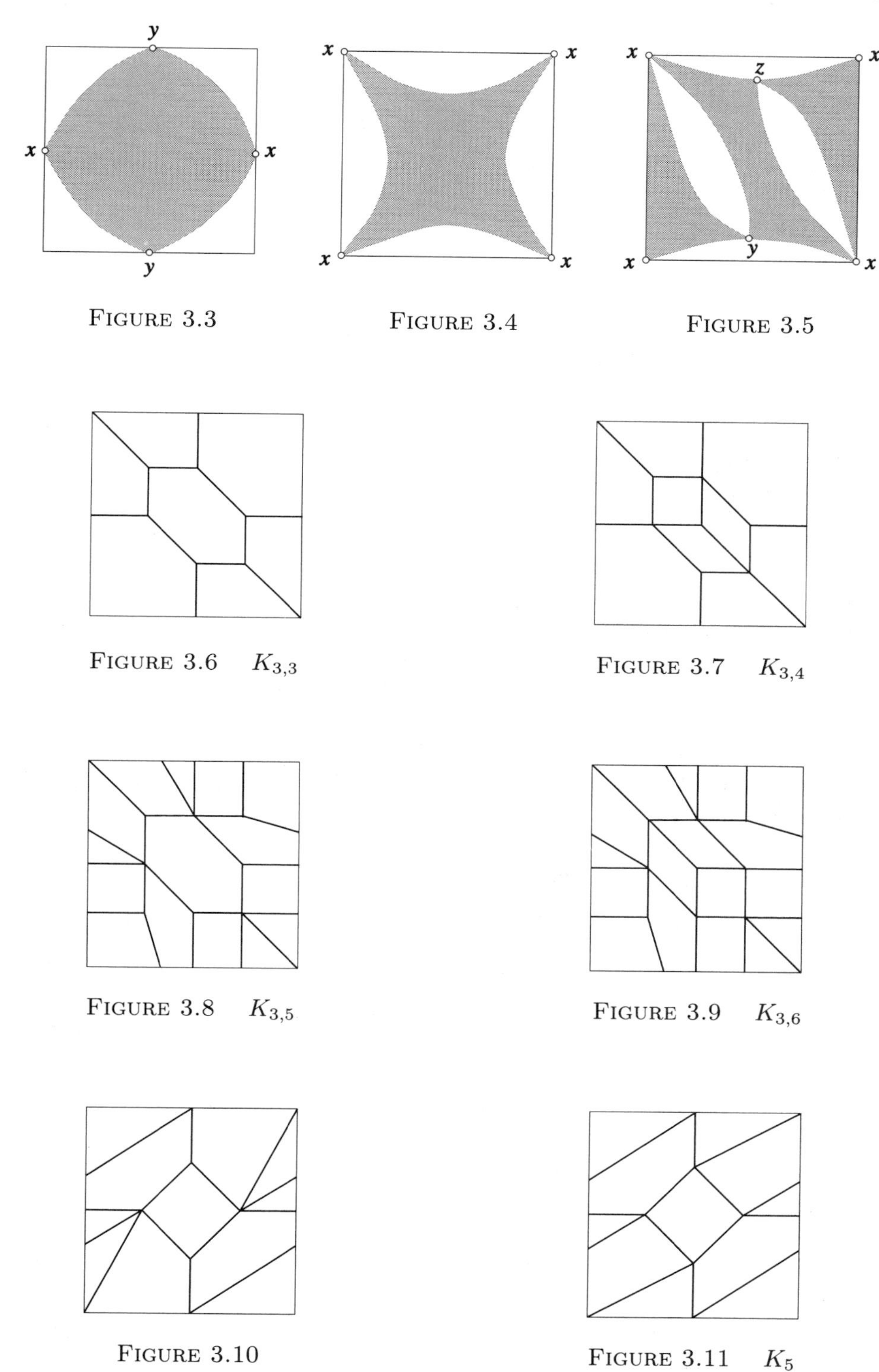

FIGURE 3.3

FIGURE 3.4

FIGURE 3.5

FIGURE 3.6 $K_{3,3}$

FIGURE 3.7 $K_{3,4}$

FIGURE 3.8 $K_{3,5}$

FIGURE 3.9 $K_{3,6}$

FIGURE 3.10

FIGURE 3.11 K_5

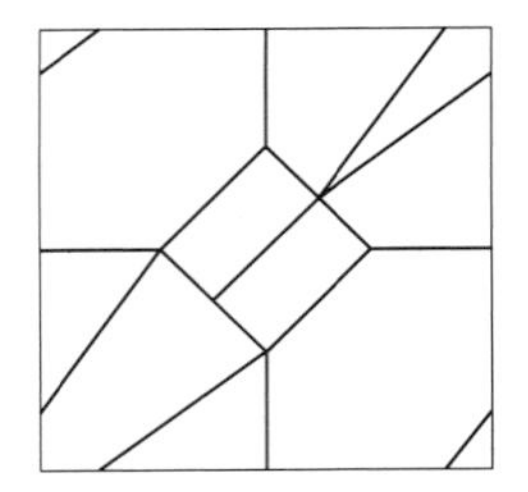

FIGURE 3.12

THEOREM 3.2. *Let G be a toroidal map with representativity $\rho(G) \geq 2$. If G is 3-reduced, then it contains no two disjoint essential cycles if and only if either $G = K_5$ embedded as in Figure 3.11, or there are distinct vertices $x, y, z \in V(G)$ which cover all the edges and so that between any two vertices $u \in \{x, y, z\}$ and $v \in V(G) \backslash \{x, y, z\}$ there are no parallel edges.*

Note. The actual cases coming out of Theorem 3.2 are depicted in Figures 3.6–3.12. We should add all their submaps (having $\rho = 2$). Note that an edge deletion in Figures 3.6–3.9 gives rise (after a 2-reduction) to an edge between x, y, z.

It is worth mentioning the similarity of the obtained characterization with the Dirac's graphs containing no two disjoint cycles. The case of Figure 3.1 corresponds to forests (graphs without cycles, vs. maps without essential cycles), Figures 3.2 and 3.4 imitate graphs with a vertex whose removal yields a forest, Figure 3.3 is an analogy of the wheel, in Figure 3.11 we have a K_5, and all the other cases have the property that there is a set of 3 "vertices" whose removal leaves a "trivial graph".

4. Planar graphs in the torus

The following corollary to the results of Section 3 is needed in [MR].

THEOREM 4.1. *Let G be a planar graph embedded in the torus with representativity $\rho(G) \geq 2$. Then G contains no two disjoint essential cycles if and only if there is a sequence of 0-, 1-, 2-, and 3-reductions transforming G into the map in Figure 3.10.*

Proof. By Lemma 2.1, reductions preserve the representativity. Therefore we may use Theorem 3.2 and its proof in Section 6 which shows that the only 3-reduced map with $\rho \geq 2$ and no two disjoint essential cycles which does not contain $K_{3,3}$ or K_5 is the map on Figure 3.10. □

The reductions can not increase the degrees of vertices (except when introducing a new vertex of degree 3). Therefore the only non-trivial 3-reductions leading to the map in Figure 3.10 must have been performed in disks around the vertices of degree 3. Theorem 4.1 therefore clearly describes the structure of arbitrary planar graphs in the torus with representativity 2 and without two disjoint essential cycles.

5. Proof of Theorem 3.1

Let G be a toroidal map containing no two disjoint essential cycles, and assume that $\rho(G) \leq 1$. The case when $\rho(G) = 0$ is easy and we leave the details to the reader. So we assume now that $\rho(G) = 1$. Then there is an essential curve γ in the torus such that $cr(\gamma, G) = 1$. We may assume that γ intersects G at a vertex, say $x \in V(G)$. Cutting the torus along γ and redrawing the map, so that γ corresponds to the bottom and the top sides of the torus under the usual representation, we get the structure as in Figure 5.1.

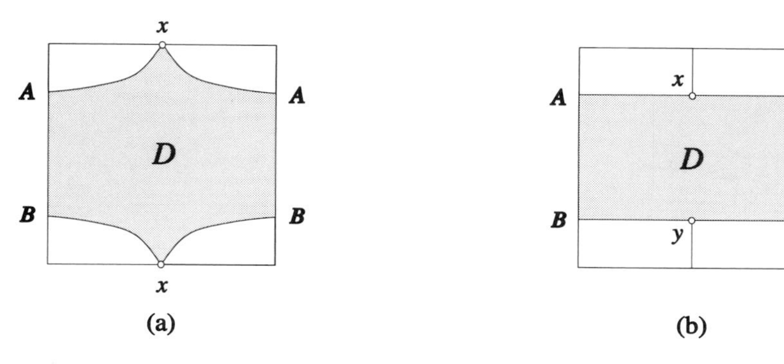

FIGURE 5.1

Let F be the face of G containing γ. Denote by A and B, respectively, the part of the facial walk of F lying in the upper and in the lower part of the drawing, respectively. Since $\rho(G) > 0$, these are well-defined. It is easy to see that we may assume that G is 2-reduced. Then A and B are essential cycles of G. Consider first the case of Figure 5.1(a). If, besides x, they have another vertex y in common, we have the structure of Figure 3.3. Thus we may assume that $A \cap B = \{x\}$. Consider the faces in D containing x at A on the boundary. Take the modulo 2 sum of the edges of A together with the boundaries of these faces. The obtained Eulerian graph is homologic to A, so it contains an essential cycle C. If a face in D contains x at A and at B, we have the structure of Figure 3.4 (after a suitable re-drawing). Therefore we may assume that $x \notin C$. Since $B \cap C \neq \emptyset$, there is a vertex $y \in B$ lying in a face of D together with x at A. A similar argument, used from "below", shows that there is a face in D containing x at B and a vertex $z \in A$. Clearly, $y \neq z$, and we have Figure 3.5.

It remains to consider the case of Figure 5.1(b). In this case, A and B have a vertex in common. Whether this vertex is equal to x or not, it turns out that G fits the structure of Figure 3.3. The details are left to the reader.

Conversely, assume that G is embedded having the structure as in one of the Figures 3.1–3.5. In the first case, there are no essential cycles at all. In the second and the fourth case, each essential cycle contains x. Having Figure 3.3, each essential cycle either contains x or y (or both), but one containing x, another y, they would cross. In the last case of Figure 3.5, each essential cycle either contains x, or it contains both, y and z. But the first case prevents the existence of a disjoint cycle through y, z, and vice versa. This completes the proof. □

6. Proof of Theorem 3.2

In this section we will assume that G is a toroidal map. It is easily seen that maps satisfying the conditions of Theorem 3.2 do not have two disjoint essential cycles.

To prove the converse, we will assume that $\rho(G) \geq 2$ and that G contains no two disjoint essential cycles. We will also assume that G is 3-reduced.

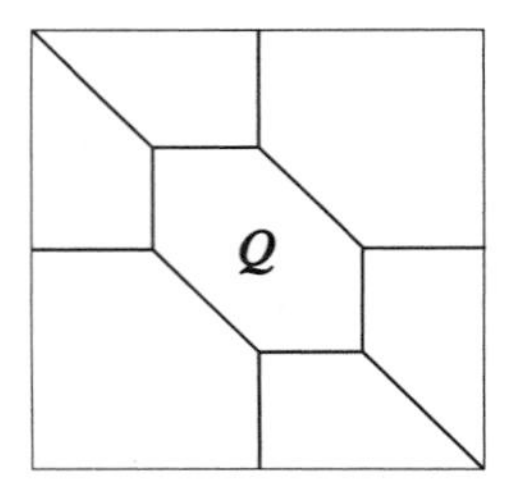

FIGURE 6.1 A $K_{3,3}$

LEMMA 6.1. *If G contains a submap homeomorphic to $K_{3,3}$ as shown on Figure 6.1, then G is a $K_{3,k}$ $(3 \leq k \leq 6)$ embedded as on Figures 3.6–3.9.*

Proof. Let K be the given submap of G homeomorphic to $K_{3,3}$. First we will prove that we may assume K has no *local bridges*, i.e., bridges attached to a single branch of K. We define

$$c(K) = |V(K)| + \sum_B |V(B) \backslash V(K)| \tag{1}$$

where the sum runs over all local bridges of K. Assume that, among all possible choices for K, we take the one with minimal $c(K)$. Suppose now that K has local bridges. Since G is 2-reduced, there must be a local bridge B which overlaps with a non-local bridge B'. Let B be attached at the branch e of K, and let p, r be its "leftmost" and the "rightmost" attachments on e, respectively. Since B' overlaps with B, it has an attachment q on e which lies between p and r. Denote by K' a submap of G obtained from K by replacing the segment from p to r on e by a path in B. Then K' is a submap homeomorphic to $K_{3,3}$. It is easy to see that $c(K') < c(K)$ since at least q does not contribute in (1) any more. This contradicts the minimality of K.

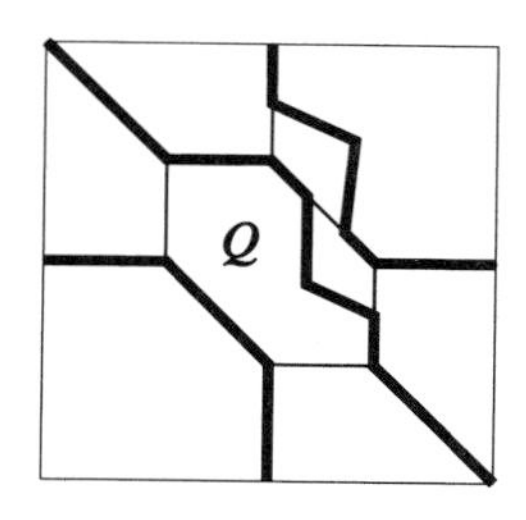

FIGURE 6.2

Let B be a bridge of K. Then the attachments of B are restricted to two adjacent branches of K (including their endvertices), or B is attached to 3 vertices of the bipartition of $K_{3,3}$. This can be seen as follows. We may assume that B is embedded in the "central" face Q. Suppose first that B is attached to an interior vertex x of a branch e of K. Then any attachment on a branch not adjacent to e, gives rise to two disjoint essential cycles in $K \cup B \subseteq G$. If a vertex of attachment x is a main vertex then, similarly, B can not be attached to the vertex y of Q opposite x, or to a vertex in an open branch at y. Since x is any vertex of attachment of B, one easily verifies that the above claim about attachments must be satisfied.

Suppose now that there is a bridge of K attached to two adjacent branches of K. Since G is 3-reduced and K contains no local bridges, there is a branch uv of K and bridges A, B of K attached as shown on Figure 6.2. Then G has two disjoint essential cycles (thick cycles in Figure 6.2). It follows that every bridge of K is attached to the main vertices only (to at least 3 of them), since the graph is 3-reduced. But the 3-attached bridges can be 3-reduced, each to a single vertex. If two such bridges are attached to different triples of main vertices of K, one easily finds two disjoint essential cycles. Finally, up to symmetries there are only four possibilities for G as exhibited in Figures 3.6 – 3.9. □

From now on we exclude the above case. We will first prove that G contains a submap K homeomorphic to K_4 shown in Figure 6.3.

LEMMA 6.2. *Let G be a toroidal map with representativity* 2 *such that there are no two disjoint essential cycles in G. If the embedding is 3-reduced then G contains a submap K homeomorphic to K_4 whose embedding is shown on Figure 6.3.*

The proof will start by a sequence of claims interlaced by introduction of notation and some small comments. In all of the claims we will assume the conditions of the Lemma and all the previous definitions and results. Moreover, we will assume that G contains no submap homeomorphic to the map in Figure 6.3.

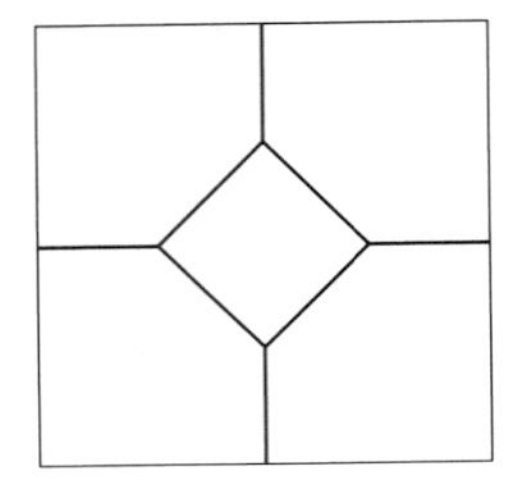

FIGURE 6.3 K_4

Since $\rho(G) \geq 2$ and G is 2-reduced, every face of G is bounded by a (simple) cycle of G. It follows, in particular, that there is a closed disk D in the torus, which is a union of (closed) faces of G, and is maximal in the sense that no other such disk properly contains D. Denote by C the boundary of D, and let $K = G \cap D$ be the subgraph of G lying in D. Clearly, C is a cycle of G.

A (closed) face of G is an *outer face* if it is not contained in D. If F is an outer face having an edge in common with D, then $F \cap D$ is not connected, since otherwise $D \cup F$ would be a disk contradicting the maximality of D. Therefore C separates the boundary cycle of F into two or more paths, $P_1, P_2, \ldots$, each of them joining two vertices on C and having no intermediate vertex on C. For each i, denote by C_i a cycle obtained from P_i and a segment on C between the endpoints of P_i (there are two choices). These cycles are called *fundamental cycles* of F with respect to D.

Claim 1. Every fundamental cycle of an outer face F, sharing an edge with D, is essential.

Proof. If it bounds a disk D', then this contradicts the maximality of D, since $D \cup D'$ is a disk. $\square$

Let $e = xu$, $f = yv$ be distinct feet of the same bridge B of K, where $x, y \in V(C)$ are vertices of attachment. Choose a path in $G - K$ joining u and v, and a segment of C joining x and y (the segment is assumed to be trivial if $x = y$). Denote by $C(e, f)$ the cycle obtained by taking the two edges e, f, the path and the segment. As in the proof of Claim 1, the maximality of D yields:

Claim 2. If $x \neq y$, then $C(e, f)$ is essential.

Claim 3. No component of $G - K$ contains an essential cycle.

Proof. Suppose that L is a component of $G - K$ containing an essential cycle S. Let F be an arbitrary outer face of G sharing an edge with D. Let C_1, C_2 be two of its fundamental cycles. Clearly, none of them can cross S since out of ∂D, C_1 and C_2 follow the boundary of a face. Therefore C_1 and C_2 are both homotopic to S. Since C_1 and C_2 are disjoint out of D, each of them can touch S only from one side, one of them from "the left", the other from "the right" if we imagine S to be "vertical". Now, any other outer face $F' \neq F$ of G sharing an edge with D would also touch S from both sides. This is now impossible since F' either lies between C_1 and S (the part not containing C_2), or between C_2 and S (the part not containing C_1). Since $\rho(G) > 1$, there are at least two appropriate faces F, F'. This gives a contradiction to the requirement that both of them touch S from both sides. □

Claim 4. Let B_1, B_2 be distinct bridges of K, and let x_i, y_i be vertices of attachment of B_i, $i = 1, 2$. Then at least two of the vertices x_1, y_1, x_2, y_2 are equal.

Proof. Assume all the attachments are distinct. For $i = 1, 2$, let e_i be a foot of B_i at x_i, and let f_i be a foot at y_i. If x_1, y_1, x_2, y_2 appear on C in that order, then it is clear (since $B_1 \neq B_2$) that $C(e_1, f_1)$ and $C(e_2, f_2)$ can be chosen to be disjoint. By Claim 2 this is not possible. Therefore we may assume that the order of attachments on C is x_1, x_2, y_1, y_2 (they interlace). However, this gives rise to a subdivision of K_4 which is embedded as shown on Figure 6.3. □

Claim 5. Every bridge of K has at least two vertices of attachment.

Proof. By Claim 3, each component L of $G - K$ is contained in an open disk D_L. We may assume that D_L contains only vertices of L and only edges of L and parts of feet of the bridge B of K containing L. We may also assume that, when a foot of B leaves D_L, it does not return to it any more.

Assume now that B has a single vertex of attachment. If for each pair e, f of feet of B, the cycle $C(e, f)$ is contractible, then there is a nontrivial 1-reduction which eliminates B. Otherwise there are feet e, f of B which are consecutive on $\partial\overline{D_L}$, according to how the edges leave D_L, and such that $C(e, f)$ is essential. The face of G containing the part of $\partial\overline{D_L}$ between e and f therefore contains an essential curve meeting G only at the attachment of B. However, this contradicts $\rho(G) \geq 2$.

The same arguments as above (only much simplified) resolve the case when a bridge is just an edge. □

Claim 6. Let D be a closed disk in the torus, and let $\gamma_1, \gamma_2, \gamma_3, \gamma_4$ be pairwise nonhomotopic essential simple closed curves in the torus. Then a pair of γ_i, γ_j $(1 \leq i < j \leq 4)$ cross each other out of D.

Proof. Any two nonhomotopic essential curves in the torus must cross at least once. Assuming that the γ_i $(i = 1, 2, 3, 4)$ do not cross out of D, they all cross each other in D. Contract D to a point x. This does not change the homotopies of the curves. It is easy to see that each γ_i gives rise to an essential simple closed curve γ_i' and a set of contractible loops. Consider now $\gamma_1'', \ldots, \gamma_4''$ which are obtained from $\gamma_1', \ldots, \gamma_4'$ (respectively) by splitting the curves at the places where they touch to get curves disjoint apart from their common point. They divide the torus into a number of regions. Since the curves are essential and pairwise nonhomotopic, each such region is bounded by at least 3 of them. A simple application of the Euler's formula now finishes the proof. The details are left to the reader. □

Claim 7. K has at least two bridges.

Proof. Assume that B is the only bridge of K. Choose an outer face F sharing an edge with D, and consider its fundamental cycles $C_1, C_2, \ldots$. Denote by $P_1, P_2, \ldots$ the corresponding paths on $\partial F \backslash E(C)$ connecting vertices of C. Since ∂F is a cycle, two paths P_i, P_j $(i < j)$ can only intersect if $j = i + 1$. In this case the initial vertex of P_j is the same as the terminal vertex of P_i. Consequently, if we have at least 3 fundamental cycles of F, then we either get a K_4 or two disjoint essential cycles. (Cf. the proof of Claim 4.) Hence we have only two fundamental cycles of F and P_1, P_2 touch. See Figure 6.4. Note that C_1 and C_2 are homotopic. The face F was chosen arbitrarily. If F' is another outer face with an edge on C, it also gives rise to a homotopic pair of fundamental cycles C_1', C_2'. Since K has only one bridge, these can not be homotopic to C_1 and C_2.

G is 3-reduced. Therefore B has at least 4 vertices of attachment, and by Figure 6.4 there are at least 4 outer faces with an edge on C. They give rise to four pairwise nonhomotopic cycles which do not cross out of D. By Claim 6 this is impossible. □

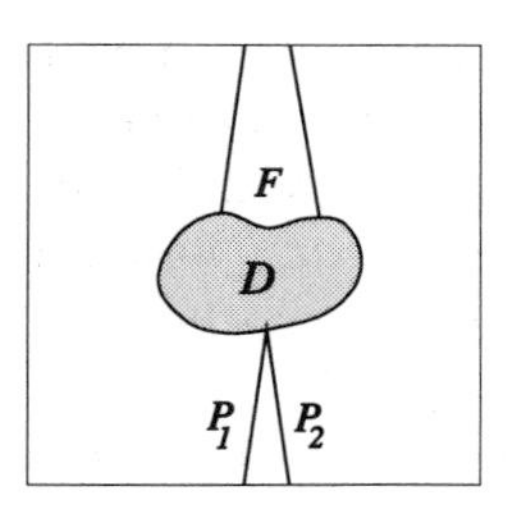

FIGURE 6.4

Claim 8. No bridge of K has more than two vertices of attachment.

Proof. If B has 4 or more vertices of attachment, any other bridge has two vertices of attachment (Claim 5) which are distinct from a pair of attachments of B. We are done by Claim 7 and Claim 4.

Suppose now that a bridge B of K has exactly three vertices of attachment. Since the total number of attachments is at least 4, there is a bridge B' with a new vertex of attachment, and in B' and B we can find pairs of distinct attachments. Again, we are done by Claim 4. □

From now on we may assume that we only have bridges of K with two attachments.

Claim 9. Every bridge of K is an edge.

Proof. A bridge B with two attachments which is not an edge would give rise to a nontrivial 2-reduction, unless it contains two feet e, f attached at the same vertex x on C such that $C(e, f)$ is essential. Now any bridge is attached to x, since otherwise it gives rise to an essential cycle disjoint from $C(e, f)$. Also, no bridge different from B can have two feet attached at a vertex $y \in V(C)$, $y \neq x$. But since $\rho(G) \geq 2$, we have $\rho(G - x) \geq 1$ (cf. [RV]). This implies that B also has a pair of feet e', f' at the other vertex of attachment y ($y \neq x$) such that $C(e', f')$ is essential. But now, by symmetry, it follows that every bridge of K is attached at y as well. This contradicts the 2-reducibility. □

Now we are well prepared to finish the proof of Lemma 6.2. The bridges of K are just edges. Any two of them have a vertex in common, since otherwise we either get a K_4 or two disjoint essential cycles, depending whether their ends on C interlace, or not, respectively. Denote by T the graph consisting of the bridges of K (and their vertices of attachment). As explained above, T does not have a 2-matching (two edges without a common vertex) and does not have a 1-cover (a vertex whose removal leaves only isolated vertices). Therefore T is not bipartite (by the König–Egervary's Theorem [B]). So T contains a cycle of odd length. Without a 2-matching this can only be a triangle. Finally, any edge of T adjacent to a fourth vertex can be extended to a 2-matching by one of the edges in the triangle. A contradiction. Lemma 6.2 is proved. □

Till the rest of this section we will assume that $\rho(G) \geq 2$, G has no 2 disjoint essential cycles, but there is a submap K of G homeomorphic to K_4 as shown in Figure 6.3. Let Q denote the quadrangular face of K. We will assume that Q is as large as possible in the sense that no other subdivision K' of K_4 has its quadrangular face Q' which properly contains Q. Denote by R the other face of K. The relative K-components embedded in R will be called *outside components* or *outside bridges* of K. Since $\rho(G) \geq 2$ there is at least one outside component. By the following lemma the attachments of outside components are quite restricted.

LEMMA 6.3. *Let B be an outside component of K. If B has a vertex of attachment to a branch or a main vertex of K at the side designated by a shading in Figure 6.5 (a), (b), or (c), respectively, then every other foot of B either attaches to the same vertex or a vertex in the part designated in Figure 6.6 (a), (b), or (c), respectively.*

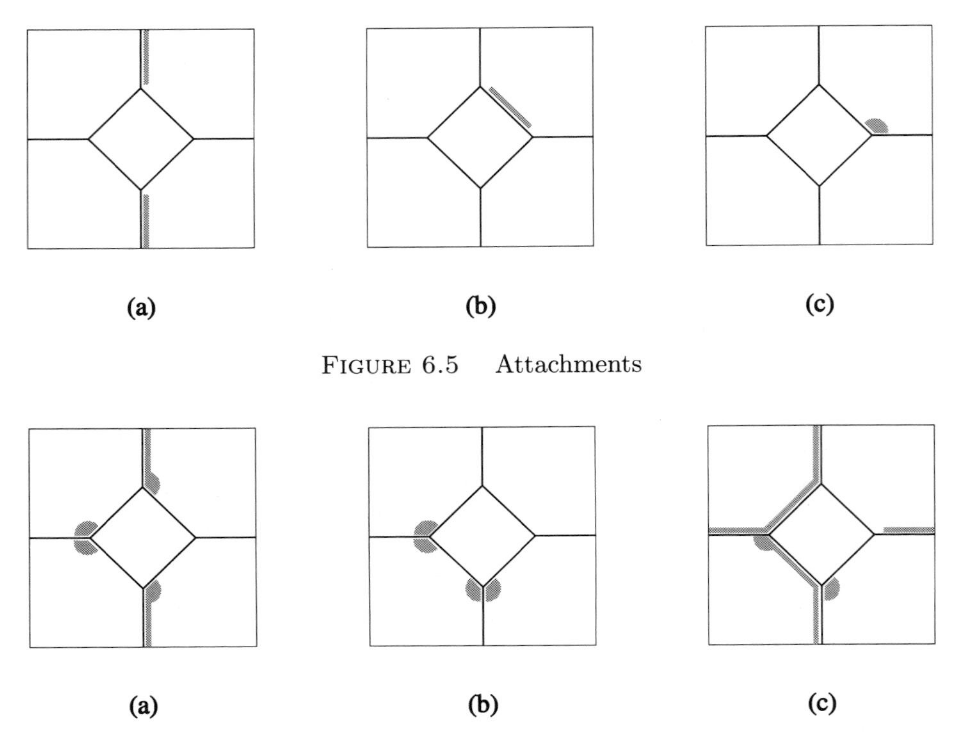

FIGURE 6.5 Attachments

FIGURE 6.6 Other attachments

Proof. In each case particular places of possible attachments can be excluded since B being attached at that place would either give rise to the case of Lemma 6.1, yield the existence of two disjoint essential cycles, or contradict the maximality of Q. $\square$

LEMMA 6.4. *Up to symmetries we may assume that G contains a subgraph K homeomorphic to one of the graphs in Figure 6.7 (a)–(e) such that:*

a) *the quadrangle Q of K_4 is maximal, and*

b) *every K-component is attached only to the points on the boundary of Q.*

Proof. Denote by K' the submap of G homeomorphic to K_4 as in Figure 6.3. Assume that B is its outside component which is attached at a vertex not on the boundary of Q. Denote by b the corresponding branch of K'. By Lemma 5.3 (Case (a)), B is attached only on one side of b. If B is attached only at vertices on this branch, say from the left side, then we can replace a part of b by the "leftmost" path of B. After a number of such changes we will definitely come to the case, when no K'-component is attached only to vertices of an outside branch of K' (otherwise contradicting 2-reducibility).

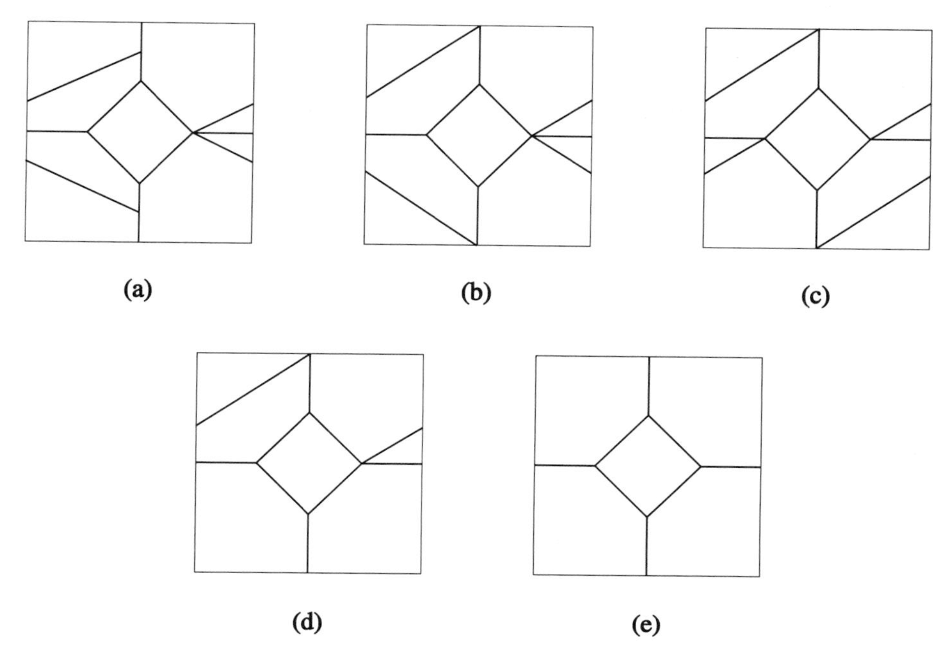

FIGURE 6.7 Subgraph K

Assume now that B is an outside K'-component attached at an interior vertex of b. Now we know that B is attached at a vertex not on b (compare with Figure 6.7(d)). Denote this vertex by x. By Lemma 6.3 (Case (c)), B is attached only at one side of x. Therefore, if B is not an edge, it is attached to b at more vertices. Let y, z be the attachments of B on b as close to each of the endpoints of b as possible. Since G is 3-reduced, there is a K'-component attached to b between y and z. Now we have a subgraph of G as shown on Figure 6.8. The thick cycles in Figure 6.8 are disjoint and essential which are assumed not to exist. Therefore B is just an edge.

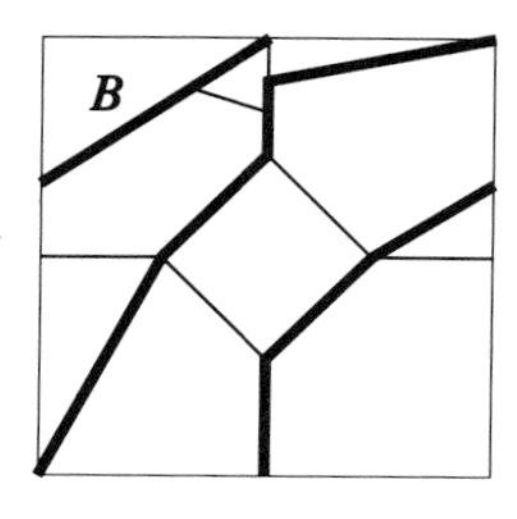

FIGURE 6.8

If two bridges are attached at different sides of b, they must share the vertex on b. Otherwise there are two disjoint essential cycles. It is now easy to see that we must have Case (c) of Figure 6.7. If two bridges of K' attach at the same side of b, then we have just two (by 3-reducibility). So we have Case (a) or (b) of Figure 6.7. Two bridges of K' attached at a point in the interior of distinct branches of K' not on ∂Q also give rise to two disjoint essential cycles. Therefore we are left with Cases (d) or (e). $\square$

LEMMA 6.5. *In Cases (a) and (b) of Figure 6.7, we have* $\rho(G) \leq 1$.

Proof. Denote the main vertices on the boundary of Q by a, b, c, d, respectively, so that a is the vertex of degree 5 in K. Consider the face of K which has repeated vertices on its boundary. Note that a appears twice, and between the two appearances of a there are only branches of K without any attached relative K-components. This implies that $\rho(G) \leq 1$. $\square$

LEMMA 6.6. *Case (c) of Figure 6.7 gives rise to two graphs on 5 vertices shown in Figures 6.9 and 6.10.*

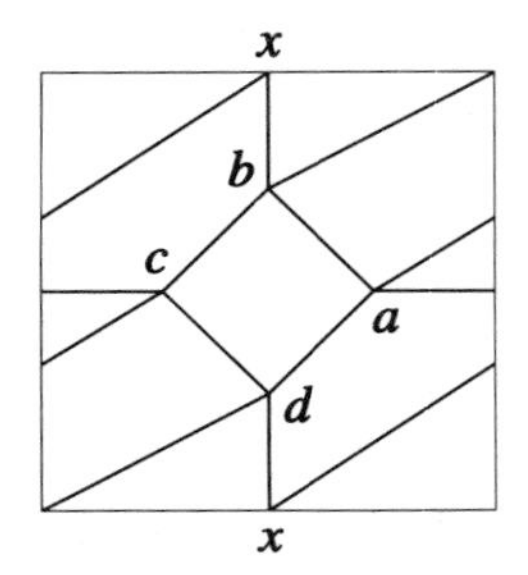

FIGURE 6.9

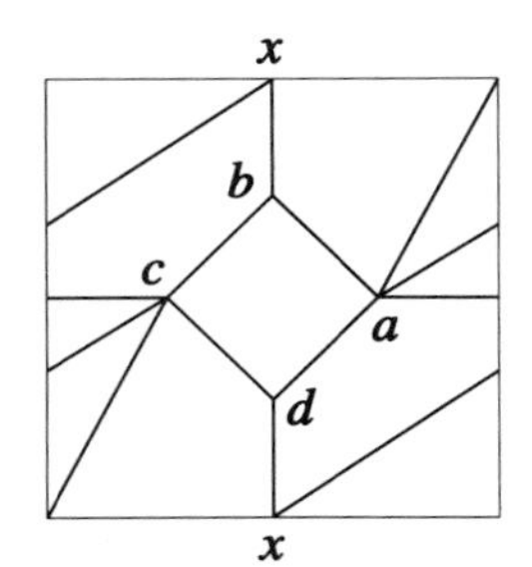

FIGURE 6.10

Proof. Denote the main vertices on the boundary of Q by a, b, c, d, respectively, so that a is on the right, b at the top. Denote by x the main vertex of K out of Q. Consider the face R of K with the facial walk $abxcdx$. Since $\rho(G) \geq 2$, there is a K-component in R. Every such component is just an edge (by the maximality of Q and 2-reducibility). If it attaches neither to a nor to d, we find two disjoint essential cycles. Similarly we see that every such K-component must attach either to b or c. Therefore we have two possibilities which are depicted in Figures 6.9 and 6.10. In none of the cases there can be additional bridges in R. It is also easy to verify that the two 4-gons $adxc$ and $cbxa$ of K contain no bridges of K (they would give rise to a 2-reduction or two disjoint essential cycles).

It remains to show that no bridge of K is in Q. In case of Figure 6.9 we have the essential cycle xbd. A K-bridge in Q overlapping with bd then gives rise to a disjoint essential cycle. By Lemma 2.2, there is a face A in Q containing b and d on its boundary. Similarly we see that there is a face B containing a and c. Clearly, $A = B$. Now, by the 2-reducibility we have $A = B = Q$.

In case of Figure 6.10 the cycle ac (of length two) is essential. As above this implies that there is a face in Q containing a and c. By the 3-reducibility, it follows that Q contains no bridges of K. $\square$

The arguments used in the above proof to show that there are no bridges of K in Q will be repeatedly used later. Let us thus state this as a lemma.

LEMMA 6.7. *Suppose that S_1, S_2, S_3, S_4 are disjoint segments on the boundary of Q, appearing in the given order. If G contains an essential cycle C, whose intersection with Q is contained in $S_1 \cup S_3$, and there is a path out of Q and disjoint from C joining a vertex of S_2 with a vertex of S_4, then in Q there is a face containing a vertex of S_1 and a vertex of S_3.*

LEMMA 6.8. *In Case (d) of Figure 6.7, G contains vertices p, q, r such that every bridge of $\{p, q, r\}$ is just an edge, or a vertex attached with one edge to each of p, q, r. In each case, G either contains a subgraph isomorphic to $K_{3,3}$, or G is the map of Figure 6.10.*

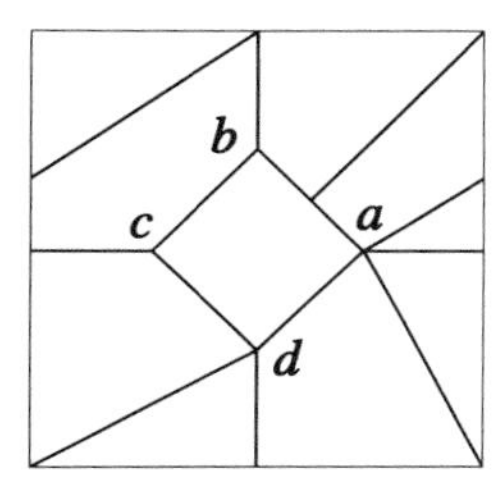

FIGURE 6.11

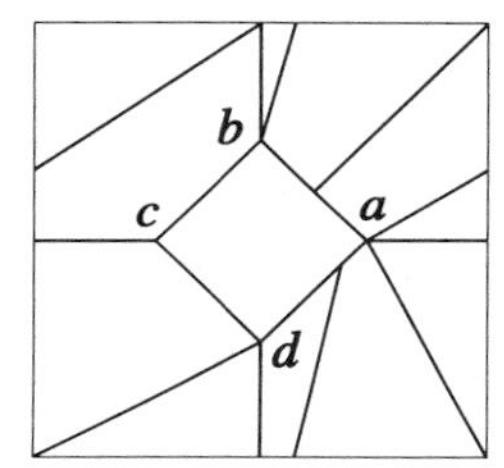

FIGURE 6.12

Proof. Under the same notation as in the proof of Lemma 6.6 consider the face R bounded by the walk $abxdacdx$. Since $\rho(G) \geq 2$ and the map is 1-reduced, the faces of G have no vertices repeated on their boundary. Therefore there must be bridges of K in R. The possible attachments in R are the branches ab, da, and cd of K. A bridge connecting cd with ad together with the cycle xba gives rise to two disjoint essential cycles unless the only attachment on ad is the vertex a. Similarly, a bridge from cd to ba must either have d as its only attachment on cd, or have a as the only attachment on ba. (Cf. the cycle dax). By Lemma 2.2, we have in R a bridge of K overlapping on ∂R with both appearances of a in order that $\rho(G) \geq 2$. By the restrictions obtained above, the only such possibility is a bridge attached at d on the branch cd, and attached at a vertex y on ab, where $y \neq a$. If the same bridge has another vertex of attachment, it is not on ab (by the maximality of Q), so it is the vertex a on the branch ad. If this bridge has only two vertices of attachment, then there must be another bridge of K in R which overlaps with both appearances of d. The only possibility for such a bridge is that it is attached at b on ab and between d and a on da (but not at d). Both possibilities are represented on Figures 6.11 and 6.12, respectively. In the second case we may as well assume that every outside bridge of K is just an edge.

Consider first the case of Figure 6.11. If $y \neq b$, then we have two disjoint essential cycles (Lemma 6.3 (b)). So $y = b$. Therefore this bridge has a, d, and y as the only attachments. It follows by the 3-reducibility that the bridge is trivial — just a vertex z together with its attachments. Consider now the subgraph H of G consisting of vertices a, b, d together with the two bridges of $\{a, b, d\}$ containing x and z, respectively. The embedding of H is cellular. Any bridge of H is attached to H at vertices a, b, d only (not necessarily all three). For each vertex $t \in \{a, b, d\}, H$ contains an essential cycle which is not using t. Therefore any bridge B of H attaches to t from one side only. Since G is 3-reduced, B is either an edge joining two of a, b, d, or a vertex of degree 3 adjacent to a, b, and d. Note that $\{a, b, d\}$ have at least 3 non-edge bridges, containing z, x, and c, respectively, so $K_{3,3} \subseteq G$.

Suppose now that we have the case of Figure 6.12. Denote by y and z the attachments on the branches ab and da, respectively. If $z \neq a$ and $y \neq b$ then the boundary of Q together with the outside diagonals ac, yd, and bz give the case settled by Lemma 6.1. Therefore either $z = a$, or $y = b$.

Assume first that $z = a$, but $y \neq b$. Since the cycle axb is essential we may use Lemma 6.7 with $S_1 = \{a\}$, $S_2 = \{y\}$, $S_3 = \{b\}$, $S_4 = \{d\}$ to see that in Q there is a face containing a and b. By the 3-reducibility it turns out now that the connected component of $G - \{a, b, d\}$ containing y is just the vertex y itself, and it is attached to a, b, d with 3 edges. Let H be the subgraph of G containing a, b, d together with the vertices x, y and their attachments to $\{a, b, d\}$ and together with the edge $bz = ba$. We conclude in the same way as above in case of Figure 6.11.

Next we consider the case $z \neq a$ and $y = b$. We assume that z is as close as possible to a. Then, if there is an outside bridge of K attached between z and a, its other attachment is on the branch cd. But this way we get an essential cycle disjoint from the cycle axb, unless the attachment is the vertex a. It follows that every outside bridge of K which is attached at the branch cd either goes to a or to b. We claim that the component of $G - \{a, b, d\}$ containing the vertex c is trivial and attached to a, b, d by 3 edges. This is evident because of the 3-reducibility if we show that in Q there is a face containing b and d. But the existence of such a face is guaranteed by Lemma 6.7 (cycle bdx, $S_1 = \{b\}$, $S_2 = \{c\}$, $S_3 = \{d\}$, $S_4 = \{a\}$).

Let H be the subgraph of G on the vertices a, b, d, x, c and with edges bd and the attachments of c, x to $\{a, b, d\}$. We conclude in the same way as in the first two cases.

In the remaining case, when $z = a$ and $y = b$, we may take for H the graph on vertices a, b, d, x and with edges bd, ab (the possibility going across!), ax, bx, dx. As above we see that there is a face in Q containing b and d, and thereafter we see that there are no additional bridges of K. The obtained map is equivalent to the map of Figure 6.10. The equivalence is realized by the permutation $(a)(bc)(dx)$. □

LEMMA 6.9. *In Case (e) of Figure 6.7, the outside bridges of K which are attached only to the main vertices of K do not give rise to a 2-representative embedding.*

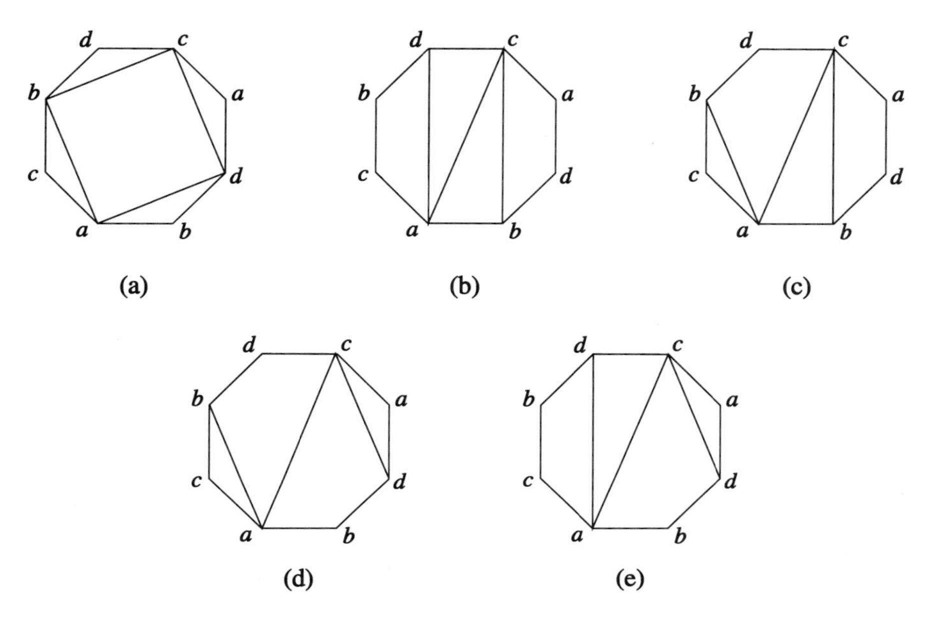

FIGURE 6.13

Proof. Consider the outer face F in Figure 6.7 (e). The outside bridges of K are just edges. In order to get $\rho(G) \geq 2$, we need for each double occurrence of a vertex on the boundary of F an edge in F overlapping with the two occurrences (Lemma 2.2). Assume (by symmetry) that the vertex a at the branch ab has the largest number of outside bridges. It is easy to see that the possible *minimal* sets of outside bridges yielding $\rho(G) \geq 2$ are the ones shown in Figure 6.13 if we additionally assume the 3-reducibility which excludes any triangles in F which do not contain an edge from the boundary of Q. In each of the Cases (a), (b), and (d), we have a pair of edges xy, wz such that $\{x, y, w, z\} = \{a, b, c, d\}$. Such a pair gives rise to two disjoint essential cycles.

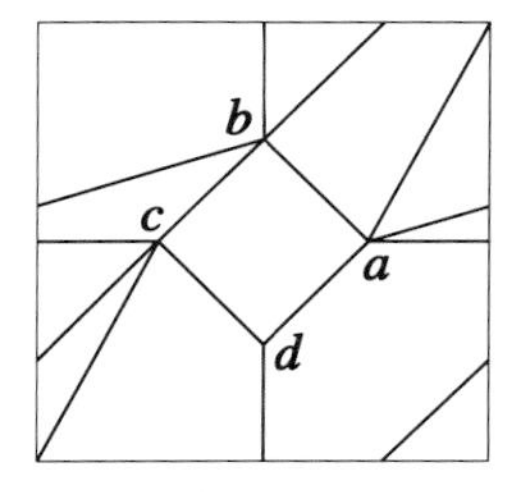

FIGURE 6.14

Case (e) is centrally symmetric to Case (c). Thus it remains to consider Case (c) which is exhibited in Figure 6.14. There is an essential cycle of length two through vertices a and c. By Lemma 6.7 ($S_1 = \{a\}$, $S_2 = \{b\}$, $S_3 = \{c\}$, $S_4 = \{d\}$), there is a face S in Q containing a and c. It follows by 3-reducibility that nothing is attached to branches ad and cd including attachments at c and a if these are coming from a face containing d. Consider now the face in Figure 6.14 bounded by the triangle abc (with the branch bc on ∂Q). It follows by the 3-reducibility that a bridge of K in Q is attached to an interior vertex of the branch bc and to a vertex $x \neq b$ on ab. Using the face S, we see that there is a disk D such that $G \cap \partial D = \{a, b, c\}$ and D contains the branches ab, bc of K and the bridge obtained above. Clearly, D gives rise to a nontrivial 3-reduction. This is a contradiction. $\square$

By Lemma 6.9, in Case (e) of Figure 6.7 there is an outside bridge of K attached at an interior vertex of a branch.

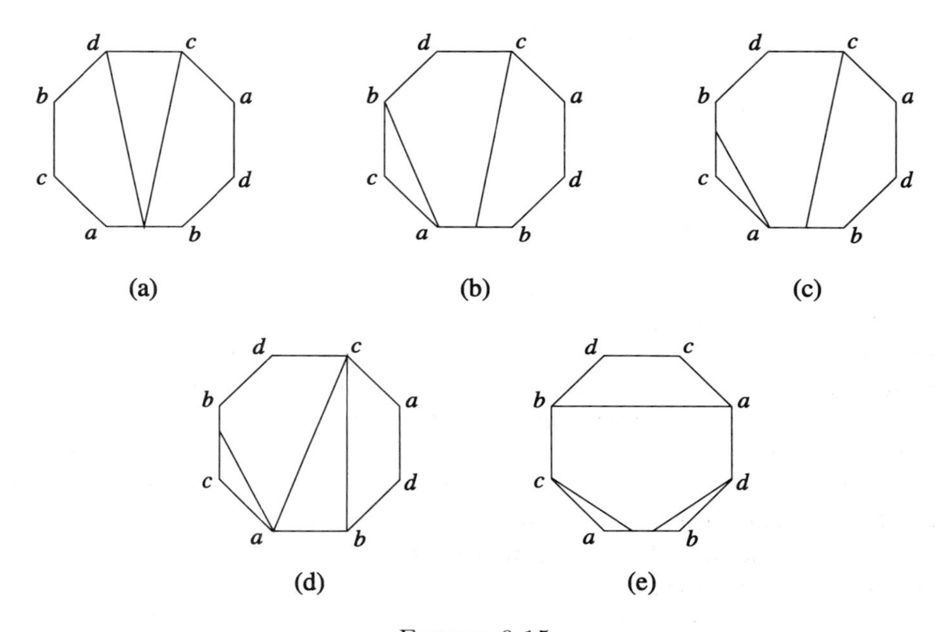

FIGURE 6.15

LEMMA 6.10. *In the remaining case only the graph $K_{3,3}$ with two additional edges embedded as shown on Figure 3.12 is obtained.*

Proof. Till the end of this section we will assume that we have Case (e) of Figure 6.7 and any minimal set of outside bridges giving $\rho \geq 2$ must contain a bridge attached to only one main vertex of K. Moreover, we know that every outside bridge of K is just an edge. Suppose that such a bridge is attached at the vertex x in the interior of the branch ab. A minimal edge-set giving $\rho \geq 2$ must separate all the double occurrences of vertices on the boundary of the outer face R of K_4. It is easy to get all such sets by exhibiting the appropriate possibilities, and having in mind that most types of bridges in R are forbidden now. Some of the obtained configurations give rise to two disjoint essential cycles. The remaining ones are collected in Figure 6.15 (a)–(e).

Cases (c), (d), and (e) are not minimal in the sense that we may delete the edge ac or bd of K_4 and still have $\rho \geq 2$ with another copy of K_4 sitting in there. These cases therefore arise from the others (Cases (a) and (b)). Case (b) can be turned into Case (a) by exchanging the edge ac of K_4 with the edge xc.

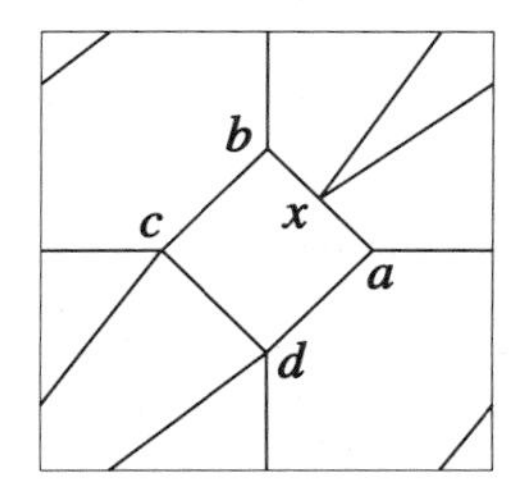

FIGURE 6.16

It remains to consider Case (a) of Figure 6.15. It is re-drawn in Figure 6.16.

By Lemma 6.7 ($S_1 = [x, b]$, $S_2 = \{c\}$, $S_3 = \{d\}$, $S_4 = \{a\}$, and the cycle $bxdb$), we see that in Q there is a face F_1 of G containing d and a vertex of the segment $[x, b]$ of ∂Q. Similarly ($S_1 = [a, x]$, $S_2 = \{b\}$, $S_3 = \{c\}$, $S_4 = \{d\}$, and the cycle $cxac$), we see that in Q there is a face F_2 of G containing c and a vertex of the segment $[a, x]$. If $F_1 = F_2$, then this face contains c and d, and there is a non-trivial 3-reduction reducing the branch cd and the edges cx, dx. Consequently, $F_1 \neq F_2$. This implies that ∂F_1 contains d and x, and ∂F_2 contains c and x. Note that no bridge of K is in the face of Figure 6.16 bounded by the triangle cxd (we would have a submap on Figure 6.1 or a non-trivial 2-reduction). Since c and d are not on the boundary of the same face contained in Q (a 3-reduction of the triangle cxd), there is a bridge of K in Q from a vertex y in the interior of the branch cd to x. By 3-reducibility, this is trivial — the vertex y is adjacent in G to c, d, and x.

So far we have shown that we have a submap represented in Figure 3.12. We need to show that there are no additional bridges of K. It suffices to see that there are no additional outside bridges since in this case any bridges in Q give rise to a non-trivial 3-reduction. The outside bridges may only be attached to the following segments on ∂Q: $[a, x]$, $[x, b]$, $[b, c]$, or $[d, a]$. By symmetry, we may consider a bridge B (if there is one) in the face bounded by $adbxc$. B is just an edge. If B is attached at the vertex $t \neq d$ on the branch da, the other end of B can not be on the segment $[x, b]$ of ∂Q (we get two disjoint essential cycles), so the other end is c. A bridge attached at d can have c or a vertex on $[x, b]$ as the other end. By symmetry, the same restrictions apply in the other face bounded by *bcaxd*. But if there is any such bridge, we have a non-trivial 3-reduction. This completes the proof. □

With the proof of Lemma 6.10 we exhibited all possible cases and we established the validity of Theorem 3.2.

References

[B] C. Berge, *Graphs*, North-Holland, Amsterdam, 1985.

[D] G. A. Dirac, *Some results concerning the structure of graphs*, Canad. Math. Bull. **6** (1963), 183–210.

[L] L. Lovász, *On graphs not containing independent circuits*, Mat. Lapok **16** (1965), 289–299.

[MR] B. Mohar, N. Robertson, *Planar graphs on nonplanar surfaces. II. The torus and the Klein bottle*, in preparation.

[RV] N. Robertson, R. Vitray, *Representativity of surface embeddings*, in "*Paths, Flows, and VLSI–Layout*," Eds. B. Korte et al., Springer, Berlin, 1990, pp. 293–328.

[S1] A. Schrijver, *Graphs on the torus and the geometry of numbers*, submitted.

[S2] A. Schrijver, *Disjoint circuits of prescribed homotopies in a graph on a compact surface*, J. Combin. Theory, Ser. B **51** (1991), 127–159.

[V] H.–J. Voss, *Cycles and Bridges in Graphs*, Kluwer Academic Publ., Dordrecht, 1991.

Department of Mathematics, University of Ljubljana, Jadranska 19, 61111 Ljubljana, Slovenia

E-mail address: bojan.mohar@uni-lj.ac.mail.si

Department of Mathematics, The Ohio State University, 231 West Eighteenth Avenue, Columbus, Ohio 43210

E-mail address: robertso@function.mps.ohio-state.edu

DIMACS Series in Discrete Mathematics
and Theoretical Computer Science
Volume 9, 1993

Layout of Rooted Trees

JÁNOS PACH AND JENŐ TÖRŐCSIK

September 16, 1992

ABSTRACT. Let S be a set of n points in the plane in general position. The depth of a point $p \in S$ is the minimum number of elements of S in a closed halfplane containing p. We prove that, if p is not the deepest point of S or the depth of p is at most $\frac{n}{3}+1$, then any tree with n vertices and with root r can be straight-line embedded on S so that r is mapped onto p. This gives a partial answer to a problem raised by Micha Perles.

1. Introduction

Let S be a set of n points in the plane in general position, i.e., no 3 of them are on the same line. We say that a graph $G = (V, E)$ with n vertices can be *laid down* (or can be *straight-line embedded*) onto S, if there exists a one-to-one mapping $\phi : V \to S$ that takes the edges of G into non-crossing straight-line segments, i.e.,

$$\big(\phi(u_1), \phi(v_1)\big) \cap \big(\phi(u_2), \phi(v_2)\big) = \emptyset \text{ for any } u_1v_1 \neq u_2v_2 \in E.$$

It is easy to see that any tree T (and, in fact, any outerplanar graph) can be laid down onto any set S with the same number of points (see for example [**2**], [**3**]). Micha Perles [**6**] raised the question whether one can arbitrarily specify the image of the root under such an embedding. The aim of this note is to give a partial answer to this question.

1991 *Mathematics Subject Classification.* 05C35.

Key words and phrases. Planar graph, tree, straight-line embedding.

Research supported by Hungarian National Foundation for Scientific Research Grant OTKA-1412 and NSF Grant CCR-89-01484

This paper is in final form and no version of it will be submitted for publication elsewhere.

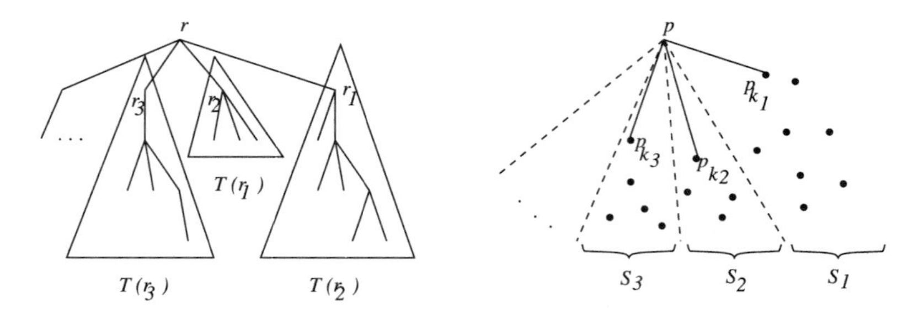

FIGURE 1.

2. Depth in Configurations

The *depth* of an element $p \in S$ is defined as the minimum number of elements of S in a closed halfplane containing p. A point $p \in S$ is a vertex of the convex hull if and only if its depth $d(p) = 1$.

THEOREM 2.1. *Let T be a tree with n vertices and with root r, and let S be a set with n points in the plane in general position. Suppose that some point $p \in S$ satisfies at least one of the following conditions:*

(i) *p is not the unique deepest point of S, or*
(ii) *the depth of p, $d(p) \leq \frac{n}{3} + 1$.*

Then there is a straight-line embedding ϕ of T onto S such that $\phi(r) = p$.

PROOF. For any point x of T, let $v^0(x) = x, v^1(x), \ldots, v^k(x) = r$ denote the vertices of the path connecting x to r in T. $v^1(x)$ is called the *father* of x, and x is the *son* of $v^1(x)$. The set of all vertices x for which the path connecting x to r passes through y induces a subtree $T(y) \subseteq T$. The vertex y is called the *root* of $T(y)$.

Algorithm 1. *The following trivial algorithm finds a straight-line embedding ϕ of T onto S with $\phi(r) = p$ in the special case when p is a vertex of the convex hull of S.*

Enumerate the points of $S - \{p\}$ by $p_1, p_2, \ldots, p_{n-1}$ in clockwise order around p. Let $r_1, r_2, \ldots$ denote the sons of r in T, and let $|T(r_j)|$ be the number of vertices of the subtree $T(r_j)$. (See Figure 1)

Let $S_i = \{p_k \mid \Sigma_{j<i}|T(r_j)| < k \leq \Sigma_{j\leq i}|T(r_j)|\}$, and find a point $p_{ki} \in S_i$ nearest to p (i=1,2, ...).

Construct recursively a straight-line embedding ϕ of the subtree $T(r_i)$ onto S_i with $\phi(r_i) = p_{ki}$ $(i = 1, 2, \ldots)$ and set $\phi(r) = p$. □

Algorithm 2. *Let p and q be two consecutive vertices of the convex hull of S, and let x be any vertex of T different from the root r. The following slightly modified version of Algorithm 1 enables us to construct a straight-line embedding ϕ of T onto S with $\phi(r) = p$ and $\phi(x) = q$.*

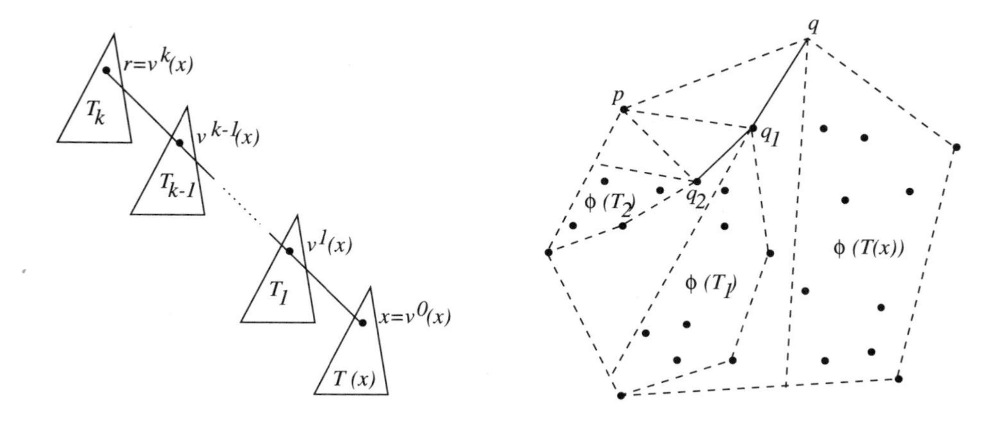

FIGURE 2.

Step 0. Let $p_1, p_2, \dots, p_{n-1}$ denote the elements of $S - \{q\}$ listed (say) in clockwise order around q, and assume by symmetry that $p_{n-1} = p$.

Use Algorithm 1 to find a straight-line embedding ϕ of $T(x)$ onto the point set $\{p_1, p_2, \dots, p_{|T(x)|-1}, q\}$, such that $\phi(x) = q$. (See Figure 2)

Let $v^0(x) = x, v^1(x), \dots, v^k(x) = r$ denote the vertices of the path connecting x to r in T.

Step i. $(1 \leq i < k)$. Let $S_i = S - \phi(T(v^{i-1}(x)))$, and let q_i be the next vertex of the convex hull of S_i that comes after p in the clockwise order. Renumber the points of $S_i - \{q_i\}$ by $p_1, p_2, \dots, p_{|S_i|-1} = p$ in clockwise order around q_i.

Use Algorithm 1 to find a straight-line embedding ϕ of $T_i = T(v^i(x)) - T(v^{i-1}(x))$ onto the point set $\{p_1, p_2, ..., p_{|T_i|-1}, q_i\}$ such that $\phi(v^i(x)) = q_i$.

Step k. Use Algorithm 1 to find a straight-line embedding ϕ of $T_k = T - T(v^{k-1}(x))$ onto S_k with $\phi(r) = p$. □

Now we are in the position to prove our main result.

Proof of Theorem. Let us build the subtree $T' \subseteq T$ from $T' = r$ by repeating the following step as long as possible.

> **If** $T - T'$ consists of at least two trees, **then** let T_{min} denote one of them having the smallest number of vertices, and
>
> **if** $|T'| + |T_{min}| \leq d(p)$, **then** set $T' = T' + T_{min}$
>
> **else** stop.
>
> **If** $T - T'$ consists of one tree, **then** let x denote its root, and
>
> **if** $|T'| + 1 \leq d(p)$, **then** set $T' = T' + x$
>
> **else** stop.

After the above process has come to an end,

> **If** $T - T'$ consists of at least two trees, **then** set $T'' = T_{min}$.
>
> **If** $T - T'$ consists of one tree, **then** set $T'' = \emptyset$.

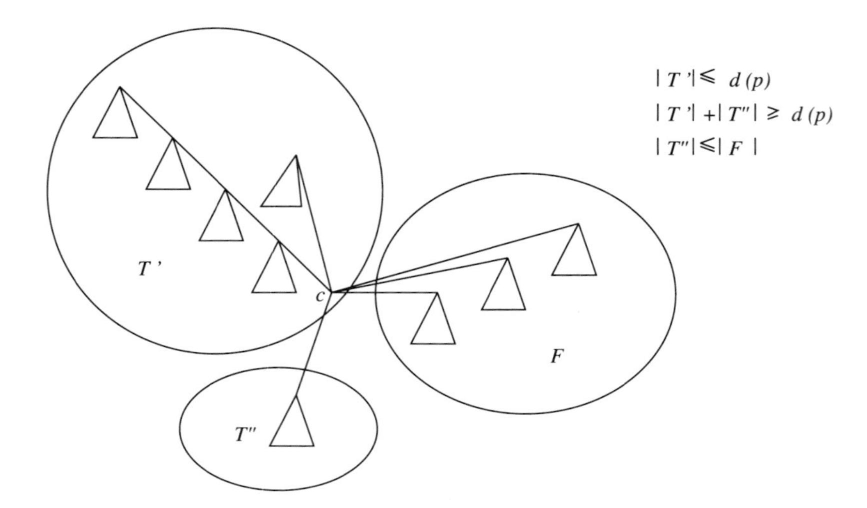

FIGURE 3.

Furthermore, let F denote the forest $T - |T'| - |T''|$ (See Figure 3).

Note that the decomposition $T = T' \cup T'' \cup F$ is usually not uniquely determined, but it can be fixed arbitrarily in the rest of the argument. It follows from the above construction that $|T'| \leq d(p)$, $|T'| + |T''| \geq d(p)$, $|T''| \leq |F|$, thus $|F| = |T| - (|T'| + |T''|) \leq n - d(p)$. Observe that T'' and each component of F are connected to the same vertex c of T', which is called the *center* of T.

Case 1. $|F| \geq d(p) - 1$.

Then $d(p) \leq |T'| + |T''| \leq n - d(p) + 1$. By the definition of $d(p)$, there exists a closed halfplane H containing p on its boundary such that $|H \cap S| = d(p)$. Letting $\overline{H}$ denote the closure of the complement of H, we have $|\overline{H} \cap S| = n - d(p) + 1$.

Suppose first that $d(p) < |T'| + |T''|$. Then by a suitable rotation of H, we obtain a closed halfplane H_{pq} with boundary line pq such that $q \in S$ and $|H_{pq} \cap S| = |T'| + |T''|$. Cut H_{pq} into two convex cones C', C'' whose apices are at q so that they have no interior points in common, $C' \cup C'' = H_{pq}$, $|C' \cap S| = |T'|$ and $|C'' \cap S| = |T''| + 1$. By Algorithm 2, we can find a straight-line embedding ϕ of T' onto $C' \cap S$ with $\phi(r) = p$ and $\phi(c) = q$. Using Algorithm 1, $T'' \cup c$ and $F \cup c$ can be laid down onto $C'' \cap S$ and $(\overline{H}_{pq} \cap S) - \{p\}$, respectively, so that c is mapped onto q (see Figure 4).

Suppose next, that $d(p) = |T'| + |T''|$. Then $T'' = \emptyset$, and F consists of a single tree whose root is denoted by c'. Rotating H around p, now we obtain a closed halfplane H_{pq} such that $q \in S$ and $|H_{pq} \cap S| = d(p) + 1 = |T'| + 1$. Using Algorithm 2, we can find a straight-line embedding ϕ of $T' \cup c'$ onto $H_{pq} \cap S$ with $\phi(r) = p$ and $\phi(c') = q$. This can be extended to a straight-line embedding of T by laying down F onto $(\overline{H}_{pq} \cap S) - \{p\}$.

Case 2. $|F| < d(p) - 1$.

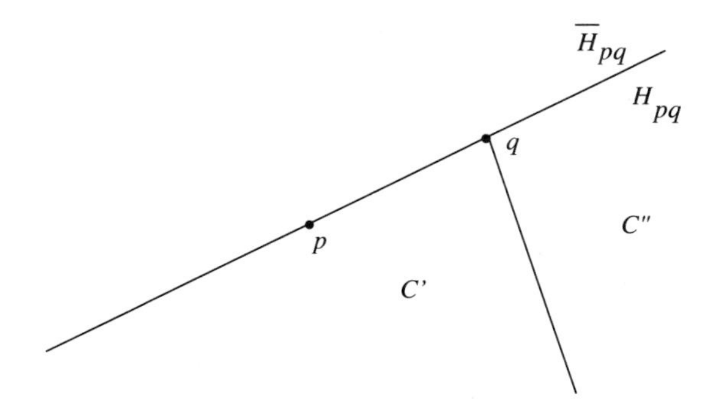

FIGURE 4.

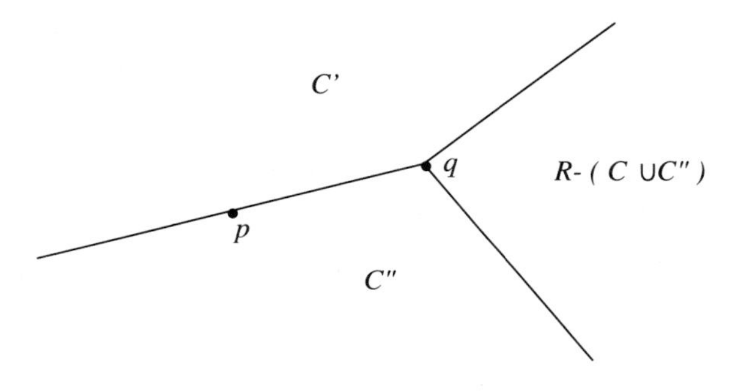

FIGURE 5.

Assume first that condition (ii) of the theorem holds, i.e., $d(p) \leq \frac{n}{3} + 1$. Then $|T''| \leq |F| \leq d(p) - 2$ and $|T'| \leq d(p)$, therefore $|T'| + |T''| + |F| \leq 3d(p) - 4 < n$, which is a contradiction.

So we can suppose that (i) is true, i.e., there exists a point $q \neq p$ in S with $d(q) \geq d(p)$. Let H_{pq} and $\overline{H}_{pq}$ denote the two closed halfplanes bounded by the line pq. Obviously, $|H_{pq} \cap S|$, $|\overline{H}_{pq} \cap S| \geq d(q)$. In view of the fact that $|T'| \leq d(p) \leq d(q)$ and $|T''| \leq |F| \leq d(p) - 2 \leq d(q) - 2$, we can find two convex cones $C' \subseteq H_{pq}$, $C'' \subseteq \overline{H}_{pq}$ whose intersection is the ray qp so that $|C' \cap S| = |T'|$ and $|C'' \cap S| = |T''| + 2$. (See Figure 5). Hence, by Algorithms 2 and 1, we can get a straight-line embedding ϕ of T' and T'' onto $C' \cap S$ and $C'' \cap S$, respectively, with $\phi(r) = p$ and $\phi(c) = q$.

On the other hand, $|(R^2 - (C \cup C'')) \cap S| = |F| \leq d(p) - 2 \leq d(q) - 2$, hence $(R^2 - (C' \cup C''))$ is either convex or it contains an open convex cone covering all points of $(R^2 - (C' \cup C'')) \cap S$. That is, ϕ can be extended to a straight-line embedding of T by laying down F onto $(R^2 - (C' \cup C'')) \cap S$. This completes the proof. □

An immediate consequence of our theorem is the following.

COROLLARY 2.2. *Let T be a tree of n vertices with root r, let S be a set of n points in the plane in general position, p_1, $p_2 \in S$. Then T can be laid down* *onto S so that the image of r is either p_1 or p_2.* □

3. Remarks on the Theorem

In fact, the above proof shows that the Theorem remains true if we replace (i) by the somewhat weaker condition that there exists $q \in S$ with $d(q) \geq d(p) - 1$.

Algorithm 2 can be generalized to establish the following statement.

PROPOSITION 3.1. *Let T be a tree on n vertices, and let $v_1, v_2, ..., v_k$ be a simple path in T. Let S be a set of n points in the plane in general position, and let $p_1, p_2, \ldots, p_j$ $(j \leq k)$ be consecutive vertices of the convex hull of S. Then for any $1 = i_1 < i_2 < \cdots < i_j = k$, there is a straight-line embedding ϕ of T onto S such that $\phi(v_{i_1}) = p_1, \ldots, \phi(v_{i_j}) = p_j$.*

4. Time Complexity

Algorithm 1 requires $O(n^2)$ time, since for each node r we have to solve a selection problem among the points corresponding to the subtree rooted at the father of r.

Algorithm 2 requires $O(n^2)$ time. Observe that in *Step i* we need to find only q_i, $\{p_1, p_2, ..., p_{|T_i|-1}\}$ and a mapping of T_i onto them. So we have to solve a selection problem and we can do the mapping in $O(|T_i|^2)$ time using Algorithm 1.

To decide if condition (ii) of the Theorem holds and if it doesn't then to find a point q with $d(q) \geq d(p) - 1$ takes $O(\frac{n^{3/2} \log^2 n}{\log^* n})$ time, where $\log^* n$ denotes the iterated logarithm function. This follows from the fact that for fized k the number of k-sets in S is $O(\frac{n^{3/2}}{\log^* n})$ (see [**5**] and [**1**] for terminology), and we can find all of them spending $O(\log^2 n)$ time on each [**4**].

To decompose our tree T into $T' \cup T'' \cup F$ takes only linear time, and the suitable rotations can be done in time $O(nlogn)$.

Therefore the overall running time is $O(n^2)$. To improve the running time we should only improve Algorithm 1.

REFERENCES

1. H. Edelsbrunner, *Algorithms in Combinatorial Geometry*, Springer-Verlag, 1987.
2. H. de Fraysseix, J. Pach and R. Pollack, *How to draw a planar graph on a grid*, Combinatorica **10** (1990), 41–51.
3. P. Gritzmann, B. Mohar, J. Pach and R. Pollack, *Embedding a planar triangulation with vertices at specified points*, Amer. Math. Monthly **98** (1991), 165–166.
4. M. H. Overmars, J. van Leeuwen, *Maintenance of Configurations in the Plane*, J. Comput. System Sci. **23** (1981), 166–204.
5. J. Pach, W. Steiger, E. Szemerédi, *An upper bound on the number of planar k-sets*, Proc. 30th Ann. IEEE Symp. Found. Comput. Sci. (1989), 72–79.
6. M. Perles, Open problem proposed at the DIMACS Workshop on Arrangements, Rutgers University, 1990.

Hungarian Academy of Sciences and Courant Institute, New York University, New york NY 10012

E-mail address: pach@cims6.nyu.edu

Eötvös University, Budapest and Department of Computer Science, Princeton University, Princeton NJ 08544

DIMACS Series in Discrete Mathematics
and Theoretical Computer Science
Volume **9**, 1993

Vertex Degrees in Planar Graphs

DOUGLAS B. WEST AND TODD G. WILL

September 17, 1992

ABSTRACT. For a planar graph on n vertices we determine the maximum values for the following: 1) The sum of the m largest vertex degrees. 2) For $k \geq 12$, the number of vertices of degree at least k and the sum of the degrees of vertices with degree at least k. 3) For $6 \leq k \leq 11$, upper and lower bounds for the latter two values, which match for certain congruence classes of n.

1. Introduction

We consider the sum of large vertex degrees in a planar graph. One approach to this is to specify a threshold k and maximize the sum of the vertex degrees that are at least k; let $f(n,k)$ denote the maximum value of this for an n-vertex planar graph. Since $K_2 \vee P_{n-2}$ is planar, we have $f(n,k) \geq 2(n-1)$ for any fixed k as long as $n \geq k+1$. Paul Erdős and Andy Vince asked whether $f(n,k) \leq 2n$ for sufficiently large fixed k; the answer is no. For $k \geq 12$ we prove:

$$f(n,k) = \begin{cases} 2n-2 & k+1 \leq n < \frac{3}{2}k - 1 \\ 2n - 16 + 6\lfloor \frac{2n-16}{k-6} \rfloor & \frac{3}{2}k - 1 \leq n \end{cases}$$

Craig Tovey independently found examples for fixed k where $f(n,k) \geq (2 + \frac{8}{k})n$; within an additive constant the optimum is $(2 + \frac{12}{k-6})n$. Fan Chung earlier observed that the sum of $o(n)$ vertex degrees in a planar graph is bounded by $2n + o(n)$. The reason for this is that the total degree in the subgraph induced by vertices of high degree is bounded by $o(n)$, and the bipartite subgraph consisting

---

1991 *Mathematics Subject Classification.* 05C35.

Both authors' research was supported in part by NSA/MSP Grant MDA904-90-H-4011.

The second author's research was also supported by Title IX Accelerated Doctoral Fellowship in Mathematics

This paper is in final form and no version of it will be submitted for publication elsewhere.

of edges from high degree to low degree vertices has at most $2n - 4$ edges. So, exceeding $2n$ by a linear amount requires a linear number of vertices of high degree.

To obtain upper bounds we first determine the maximum sum of the m largest vertex degrees. To obtain some of the lower bounds we construct triangulations in which all vertices have degree 3, or k. The constructions therefore are related to a question posed by Jerry Griggs; what is the fewest number of vertices of degree less than k in an n-vertex triangulation? We present answers for all n when $k \geq 12$ and special congruence classes of n when $6 \leq k \leq 11$. For $k < 6$ the minimum is 0, but the maximum number of vertices of degree k is $n - 2$ if $k = 4$ or if $k = 5$ and n is even.

2. The m Largest Vertex Degrees

Given a planar graph G on n vertices, let B be a set of m vertices of G with largest degree and let $D = \sum_{v \in B} d(v)$. We obtain an upper bound on D by studying the structure of the subgraph induced by B. We use $N(v)$ to denote the set of neighbors of a vertex v and G_T to denote the subgraph of G induced by a set of vertices $T \subseteq V(G)$.

LEMMA 2.1. *If $m \geq 3$ and G is a plane graph maximizing D, then G_B is a triangulation.*

PROOF. Let $S = V(G) - B$ be the set of vertices of small degree in G. We may assume that G has no edges within S, since deleting such edges does not reduce D. Hence every face of G contains a vertex of B. If G is not connected, then we can increase D by joining vertices of B on a face bounding two components of G. Hence we may assume G is connected.

Suppose $v \in S$, and consider two rotationally-consecutive edges vx, vy at v, if $d(v) \geq 2$. The independence of S implies $x, y \in B$, and the maximality of D implies $xy \in E(G)$. Hence $G_{N(v)}$ is connected. If G_B is not connected, the connectivity of G guarantees a path of length two joining components of G_B through a vertex $v \in S$. This contradicts our conclusion that $G_{N(v)}$ is connected; we conclude that G_B is connected.

If G_B is not a triangulation, then we can find three vertices x, y, z consecutive along a face of G_B, with $xz \notin E(G)$; we may assume yz is clockwise from yx at y through this face. Let the neighbors of y in clockwise order from x to z be $x, a_1, \ldots, a_p, z$. By the independence of S and maximality of D, the neighbors of x in $\{a_i\}$ are an initial segment of $\{a_i\}$; let a_r be the last neighbor of x in the order, if any. We delete the edge $a_r y$ (if $r > 0$) and replace the edges $ya_{r+1}, \ldots, ya_p$, if any, by $xa_{r+1}, \ldots, xa_p$. We can now add xz for a net increase in D. (Note: if the removal of edges from x reduces its degree so it is no longer among the m largest, we still have contradicted the maximality of D.) We conclude that G_B has no face of length exceeding 3. $\square$

LEMMA 2.2. *Let C be a closed walk in a simple plane graph G. Let S be a set of s vertices in a region bounded by C, and let R be a specified set of $r \geq 2$ vertices on the portion of C bounding it. Then there are at most $r + 2(s-1)$ edges between R and S, and this is achievable if $s \geq 1$.*

PROOF. By induction on r. For $r = 2$, the fact that G is simple yields the desired bound $2s$. Suppose $r \geq 3$. If every vertex of S has at most two neighbors in R, then the number of edges is at most $2s < r+2(s-1)$. Hence we may assume there is a vertex $v \in S$ with $k \geq 3$ neighbors in R. Without loss of generality we may assume that $N(v) \subseteq R$. Thus the edges from v to $N(v)$ complete k closed walks with segments of C bounded by vertices of R; call these $C_1, ..., C_k$. Let s_i be the number of vertices of S in the portion of the original region bounded by C_i. Since each C_i is missing at least one element of R, we can apply induction with $R_i = C_i \cap R$ to obtain a total bound of

$$k + \sum_{i=1}^{k}(|R_i| + 2(s_i - 1)) = \sum_{i=1}^{k} |R_i| + 2\sum_{i=1}^{k} s_i - k.$$

Since $\sum_{i=1}^{k} |R_i| = |R| + k$ and $\sum_{i=1}^{k} s_i = s - 1$, this simplifies to the desired bound. If $s \geq 1$, the bound can be achieved by connecting one inside vertex to all elements of R and the remaining inside vertices to two consecutive neighbors of the first inside vertex. □

THEOREM 2.1. *The maximum of the sum of the m largest vertex degrees in an n-vertex planar graph is*

$$D(n,m) = \begin{cases} n-1 & \textit{for } m = 1 \\ 2n-2 & \textit{for } m = 2 \\ 2n-16+6m & \textit{for } 3 \leq m \leq \frac{1}{3}(n+4) \\ 3n-12+3m & \textit{for } m > \frac{1}{3}(n+4) \end{cases}$$

PROOF. If $m = 1$, then $n - 1$ is clearly an upper bound achieved by a star. For $m = 2$ the bound is still clear and can be achieved by $K_2 \vee P_{n-2}$. For $m \geq 3$, let G be a planar graph maximizing D; we know that G_B is a triangulation with $3m - 6$ edges and $2m - 4$ faces. Note that each vertex of a triangulation has degree at least three if $m \geq 4$. The question then becomes how can the remaining $n - m$ vertices be added to produce the maximum value for D. Since we add edges from these vertices only to B, they will have degree at most 3, and the vertices claimed to have the m largest degrees in fact will have the m largest degrees. We know by Lemma 2.2 that the maximum contribution due to s vertices inside any triangle is $3 + 2(s - 1)$. Therefore, D is maximized by greedily distributing one vertex per face until each face has a vertex inside, for a contribution of 3 for each such vertex, and additional vertices contribute only 2. If $n \leq 3m - 4$, then the total is $2(3m - 6) + 3(n - m) = 3n - 12 + 3m$; if $n \geq 3m-4$, then total is $2(3m-6)+3(2m-4)+2(n-3m+4) = 2n-16+6m$. □

3. The Vertex Degrees Above a Threshold

3.1. Upper Bounds. In this section we consider $f(n,k)$, the maximum possible degree sum of the vertices with degree above a threshold k. Our strategy to bound $f(n,k)$ is to first bound the maximum number of vertices of degree at least k in an n-vertex planar graph; we denote this by $m(n,k)$. Then, since the bound $D(n,m)$ obtained in the previous section is monotone in m, we obtain the bound $f(n,k) \le D(n, m(n,k))$. Returning to the planar graph $K_2 \vee P_{n-2}$, we see that $m(n,k) \ge 2$ as long as $n \ge k+1$. If $m(n,k) \ge 3$, then the sum of the three largest degrees is at least $3k$, which by the previous theorem is at most $D(n,3) = 2n+2$. This implies that the special case $m(n,k) = 2$ and $f(n,k) = 2n-2$ occurs only when $k+1 \le n \le \frac{3}{2}k - 2$. Henceforth we focus on the case $n \ge \frac{3}{2}k - 1$.

THEOREM 3.1. *If $n \ge \frac{3}{2}k-1$ and $k \ge 6$, then*

$$m(n,k) \le \begin{cases} \frac{2n-16}{k-6} & \text{if } k \ge 12 \text{ or } n \le \frac{4(k+6)}{12-k} \\ \frac{3n-12}{k-3} & \text{if } 6 \le k \le 11 \text{ and } n > \frac{4(k+6)}{12-k} \end{cases}$$

PROOF. Let $m = m(n,k)$. We first note that when $n \ge \frac{3}{2}k - 1$ it is easy to construct a planar graph with three vertices of degree k. Hence we may assume $m \ge 3$. Since these m vertices are those of largest degree, $km \le D(n,m)$. Using the bound obtained in the last section, we have

$$km \le \begin{cases} 2n-16+6m & 3 \le m \le \frac{1}{3}(n+4) \\ 3n-12+3m & m > \frac{1}{3}(n+4) \end{cases}$$

Hence whenever $3 \le m \le \frac{1}{3}(n+4)$ we have the bound $m \le \frac{2n-16}{k-6}$, and whenever $m > \frac{1}{3}(n+4)$ we have the bound $m \le \frac{3n-12}{k-3}$.

If $k \ge 12$ and $m > \frac{1}{3}(n+4)$, then the second bound yields

$$12m \le km \le 3(3m-4) - 12 + 3m = 12m - 24.$$

Hence if $k \ge 12$, then $m \le \frac{1}{3}(n+4)$ and the first bound always holds.

Henceforth suppose $6 \le k \le 11$. If $n \le \frac{4(k+6)}{12-k}$ and $m > \frac{1}{3}(n+4)$, then the second bound says $m \le \frac{3n-12}{k-3}$. Together these imply $\frac{1}{3}(n+4) < \frac{3n-12}{k-3}$, which is equivalent to $n > \frac{4(k+6)}{12-k}$ and contradicts the hypothesis. Hence we must have the first bound when $n \le \frac{4(k+6)}{12-k}$. Note also that this implies $k > \frac{12(n-2)}{n+4}$, which is at least 6 when $n \ge \frac{3}{2}k - 1$, so the bound is meaningful.

If $n > \frac{4(k+6)}{12-k}$ and $m > \frac{1}{3}(n+4)$, then the second bound $m \le \frac{3n-12}{k-3}$ applies as claimed, so suppose $m \le \frac{1}{3}(n+4)$. As noted above, $n > \frac{4(k+6)}{12-k}$ is equivalent to $\frac{1}{3}(n+4) < \frac{3n-12}{k-3}$, so in this case we obtain $m \le \frac{3n-12}{k-3}$ again. □

COROLLARY 3.1. *If $n \ge \frac{3}{2}k - 1$ and $k \ge 6$, then*

$$f(n,k) \le \begin{cases} 2n-16+6\lfloor\frac{2n-16}{k-6}\rfloor & \text{if } k \ge 12 \text{ or } \frac{3}{2}k-1 \le n \le \frac{4(k+6)}{12-k} \\ 3n-12+3\lfloor\frac{3n-12}{k-3}\rfloor & \text{if } 6 \le k \le 11 \text{ and } n > \frac{4(k+6)}{12-k} \end{cases}$$

PROOF. As was our goal, we now use the fact that $D(n, m)$ is monotone increasing in m to obtain the bound $f(n, k) \leq D(n, m(n, k))$. If $k \geq 12$ or $n \leq \frac{4(k+6)}{12-k}$, then by the last theorem $m(n, k) \leq \frac{2n-16}{k-6}$, and hence $f(n, k) \leq D(n, \frac{2n-16}{k-6})$. The hypothesis of this case also guarantees $3 \leq \frac{2n-16}{k-6} \leq \frac{1}{3}(n+4)$, so we can evaluate $D(n, \frac{2n-16}{k-6}) = 2n - 16 + 6\lfloor\frac{2n-16}{k-6}\rfloor$. This leaves the case $6 \leq k \leq 11$ and $n > \frac{4(k+6)}{12-k}$, where the last theorem gives $m \leq \frac{3n-12}{k-3}$. If $\frac{3n-12}{k-3} > \frac{1}{3}(n + 4)$, we obtain the bound claimed for T. Because $D(n, m)$ is monotonic in m, this bound on T also holds if $\frac{3n-12}{k-3} \leq \frac{1}{3}(n + 4)$. □

3.2. Lower Bounds. The constructions in this section all begin with the definition of a set B of m "big" vertices (intended to have degree above the threshold), and the description of a triangulation G_B. After this we add the remaining $n - m$ vertices to faces or edges of G_B. Adding a vertex to a face means placing it inside the face and joining it to each vertex on the face. Adding a vertex to an edge uv means placing it in a face bounded by uv and joining it to both u and v. In all cases except $k = 11$, vertices are not added to edges unless a vertex has been already been added to every face. The proof of Theorem 3 guarantees that the sum of the degrees of the m big vertices will be $D(n, m)$. For $k \geq 12$ we are able to match the upper bound on $f(n, k)$ for all n. For $k < 12$ our upper bounds split into cases depending on n. We construct matching lower bounds for large n in certain congruence classes.

We note in passing that the combination of our bounds and our constructions answers some cases of a question posed by Jerry Griggs. He asked for the minimum number of vertices of degree less than k in a planar n-vertex triangulation, which is equivalent to determining the maximum number of vertices of degree at least k. For $k \geq 12$, this maximum number is always $\lceil\frac{2n-16}{k-6}\rceil$. For $6 \leq k \leq 11$, we determine the minimum number of vertices of degree less than k for appropriate congruence classes of n. In particular, for $k = 6$ and n even, this minimum number is 4, which requires four vertices of degree 3. Griggs reports that when n is odd, there is no triangulation with four vertices of degree 3 and the rest of degree (at least) 6, as proved by Grünbaum and Motzkin [**2**], so here the answer is 5 (we omit this construction). Griggs also reports that Yan-Chyuan Lin has determined the answer within 1 for the remaining cases between $k = 7$ and $k = 11$. The remainder of this section contains constructions that prove our results by showing $f(n, k) = D(n, m(n, k))$ in various cases, except that for $k = 11$ we must also improve the upper bound slightly. For $k + 1 \leq n \leq \frac{3}{2}k - 2$ we have already shown $m(n, k) = 2$, and $f(n, k) = 2n - 2 = D(n, 2)$, so again we focus on the case $n \geq \frac{3}{2}k - 1$.

THEOREM 3.2. *If $k \geq 12$, then $f(n, k) = D(n, m(n, k))$.*

PROOF. Let $2n - 16 = m(k - 6) + r$ with $0 \leq r < k - 6$, so that $m = m(n, k)$. Let $B = \{u_1, ..., u_m\}$. The graph G_B consists of the edges $\{u_iu_j\}$ such that $|j - i| \leq 3$. There are $m - 1 + m - 2 + m - 3 = 3m - 6$ edges. Figure 1 illustrates

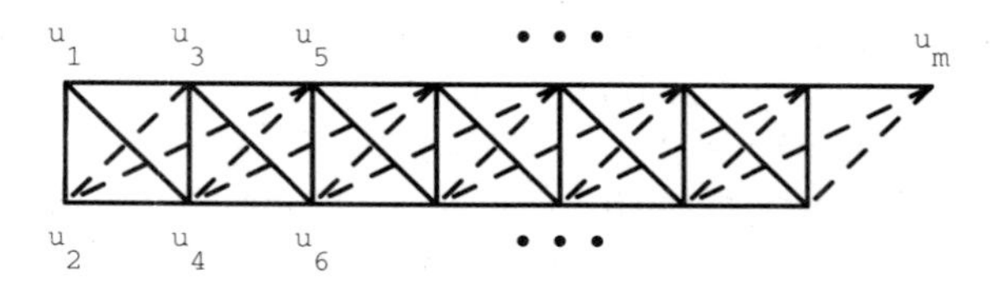

FIGURE 1.

the graph for odd m. We obtain a planar representation by drawing the dashed path around the outside.

Adding a vertex to each of the $2m-4$ faces of G_B doubles the degree of each vertex in B, giving them all degree 12 except for $d(u_1)=d(u_m)=6$, $d(u_2)=d(u_{m-1})=8$, and $d(u_3)=d(u_{m-2})=10$. To bring all vertices of B to degree 12, add 2 vertices to each of u_1u_3 and u_mu_{m-2}, and add 4 vertices to each of u_1u_2 and u_mu_{m-1}. To reach degree k, we add $k-12$ vertices to each edge $u_{2i-1}u_{2i}$ for $1 \le i \le \lfloor m/2 \rfloor - 1$. If m is even, we also add $k-12$ vertices to $u_{m-1}u_m$. If m is odd, we add $\lfloor (k-12)/2 \rfloor$ to $u_{m-2}u_{m-1}$, and we add $\lceil (k-12)/2 \rceil$ to each of $u_{m-2}u_m$ and $u_{m-1}u_m$. Altogether, the number of vertices used is $m+(2m-4)+12+(k-12)(\lfloor m/2 \rfloor)+\epsilon$, where $\epsilon=0$ if m is even and $\epsilon = \lceil (k-12)/2 \rceil$ if m is odd. Considering all cases for the parity of m and k, the formula equals $8+\lceil m(k-6)/2 \rceil$. Since $m(k-6)=2n-16-r$, we have used $n-\lfloor r/2 \rfloor$ vertices; add the remaining $\lfloor r/2 \rfloor$ vertices to the edge u_1u_2. In all cases for the parity of m and k, the construction achieves the bound $f(n,k) \le 2n-16+6m$ from the previous section. □

THEOREM 3.3. *If* $6 \le k \le 8$ *and* $n = r_kj_k + 4$, *where* $j_k = 2, 8, 10$ *for* $k=6,7,8$, *and* $r_k \ge 2,1,1$ *for* $k=6,7,8$, *then* $f(n,k)=D(n,m(n,k))$.

PROOF. For these cases our base graph G_B is a triangulation on an even number of vertices with 4 vertices of degree 4, 4 vertices of degree 5, and the remainder of degree 6. Let $m=2p$; the graphs in Figure 2 (without the dots) illustrate G_B when $m=24$. Let $B=\{a_1,\dots,a_{\lceil p/2 \rceil}\} \cup \{b_1,\dots,b_{\lfloor p/2 \rfloor}\} \cup \{c_1,\dots,c_{\lceil p/2 \rceil}\} \cup \{d_1,\dots,d_{\lfloor p/2 \rfloor}\} \cup$. The graph G_B consists of the sets $\{a_i\}$, $\{b_i\}$, $\{c_i\}$, $\{d_i\}$ inducing paths (with indices in order), the four-cycles $(a_ib_ic_id_i)$ and $(a_{i+1}b_ic_{i+1}d_i)$ for for $1 \le i \le \lfloor p/2 \rfloor$, and the two edges a_1c_1 and either $b_{p/2}d_{p/2}$ (if p is even) or $a_{\lceil p/2 \rceil}c_{\lceil p/2 \rceil}$ (if p is odd). All vertices have degree 6 except those on the cycle $(a_1b_1c_1d_1)$ of degrees 4 and 5 and those on the cycle $a_{\lceil p/2 \rceil}b_{\lfloor p/2 \rfloor}c_{\lceil p/2 \rceil}d_{\lfloor p/2 \rfloor}$ of degrees 4 and 5.

In each case, we begin with this triangulation G_B and add vertices into selected faces to achieve the bounds of the preceding section.

For $k=6$, suppose $n=2r+4$. Using the graph G_B described above for $m=2r$, add the 4 vertices into the faces indicated in Figure 2. Note that this graph is 6-regular, except for four vertices of degree 3.

For $k=7$, suppose $n=8r+4$. Begin with the graph constructed for $k=6$

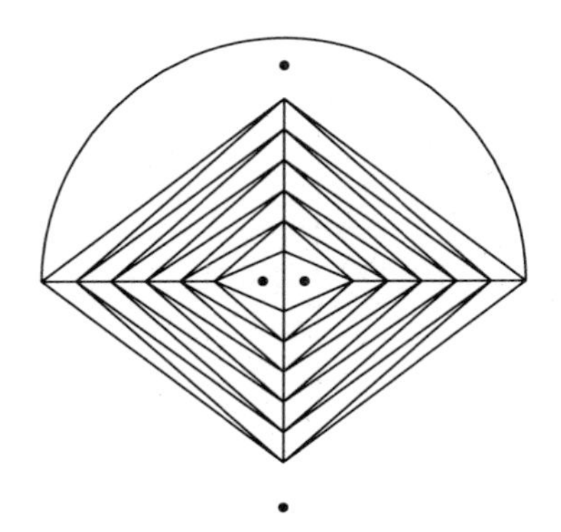

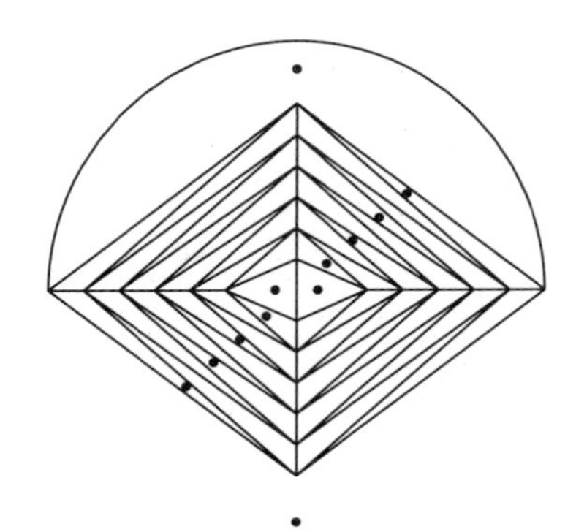

FIGURE 2.

when $m = 6r$. We increase the degree of the $6r$ vertices in B by one by adding one to each of a set of $2r$ vertex-disjoint triangles. The triangles are the partitions into consecutive triples of the sequence $a_1, b_1, a_2, b_2, a_3, b_3, \ldots$ and the sequence $c_1, d_1, c_2, d_2, c_3, d_3, \ldots$. Since m is divisible by 6, this covers the triangles exactly.

For $k = 8$, suppose $n = 10r + 4$. This is very similar to $k = 7$. To the graph constructed above for $k = 7$ and $n = 8r + 4$, add the same pattern of r additional vertices into each of the two empty wedges. That is, we place vertices in the triangles consisting of the partitions into consecutive triples of the sequence $a_1, d_1, a_2, d_2, a_3, d_3, \ldots$ and the sequence $c_1, b_1, c_2, b_2, c_3, b_3, \ldots$. $\square$

THEOREM 3.4. *If* $9 \leq k \leq 10$ *and* $n = rj_k + s_k$, *where* $j_k = 18, 21$ *and* $s_k = 10, 11$ *for* $k = 9, 10$, *and* $r \geq 1$, *then* $f(n,k) = D(n, m(n,k))$.

PROOF. For $k = 9, 10, 11$ we need a new base graph. Now G_B will be a triangulation on $m = 3p$ vertices with 12 vertices of degree 5 and the remaining vertices of degree 6. Such a graph on 30 vertices appears in Figure 3 (ignoring the dots). The construction of this graph is much like the construction of the earlier triangulation, but with six spokes instead of 4. Let the vertices be $\{a_{i,j} : 0 \leq i \leq 2, 1 \leq j \leq \lceil p/2 \rceil\}$ and $\{b_{i,j} : 0 \leq i \leq 2, 1 \leq j \leq \lfloor p/2 \rfloor\}$. The edges are the "spoke-paths" $(a_{i,1} \ldots a_{i,\lceil p/2 \rceil})$ and $(b_{i,1} \ldots b_{i,\lfloor p/2 \rfloor})$, the 6-cycles $(a_{0,j} b_{0,j} a_{1,j} b_{1,j} a_{2,j} b_{2,j})$ and $(a_{0,j+1} b_{0,j} a_{1,j+1} b_{1,j} a_{2,j+1} b_{2,j})$ for $1 \leq j \leq \lfloor p/2 \rfloor$, and the two triangles $a_{0,1} a_{1,1} a_{2,1}$ and either $b_{0,p/2} b_{1,p/2} b_{2,p/2}$ (if p is even) or $a_{0,\lceil p/2 \rceil} a_{1,\lceil p/2 \rceil} a_{\lceil p/2 \rceil}$ (if p is odd). The vertices all have degree 6, except the first and last of each spoke, which have degree 5. The four faces involving $\{a_{i,1}\}$ and $\{b_{i,1}\}$ we call the "inner" faces; similarly there are four "outer" faces involving $\{a_{i,\lceil p/2 \rceil}\}$ and $\{b_{i,\lfloor p/2 \rfloor}\}$. The remaining faces are the triangles formed by any three consecutive vertices in the six sequences $a_{i,1}, b_{i',1}, a_{i,2}, b_{i',2}, \ldots$, where $i' \equiv i \pm 1$ *mod* 3. The triangles in each such sequence form a wedge of $p - 2$ triangles.

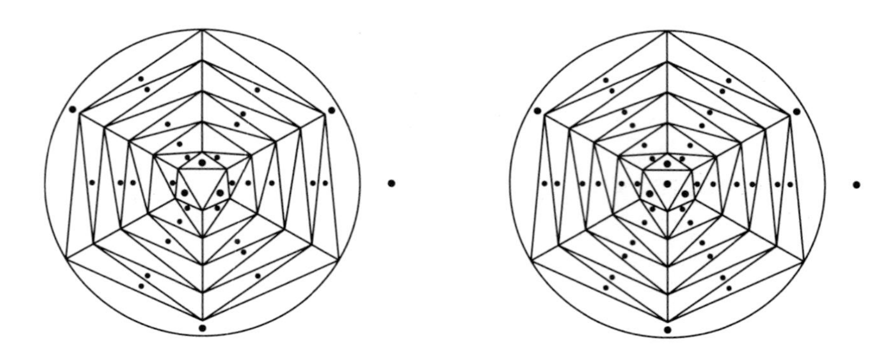

FIGURE 3.

For $k = 9$, suppose $n = 18r + 10$. Start with the graph G_B defined here for $m = 9r + 3$ vertices. Add vertices to the three inner triangles other than $a_{0,1}a_{1,1}a_{2,1}$. However, add vertices to all four outer triangles. For the successive $p - 2 = 3r - 1$ triangles in a wedge, starting from the center (low index), we describe a binary string that specifies adding a vertex to the jth triangle in the wedge if and only if the jth entry in the sequence is a 1. Note that at this point each $a_{i,1}$ has degree 7 and each $b_{i,1}$ has degree 6; however, the vertices on the outer edges of the wedges have degree 8 and those immediately before them have degree 6. In the wedge formed by the vertices of $\{a_{i,j}\}$ and $\{b_{i+1,j}\}$, we start with 11 and then alternate 010 and 011. In the wedge formed by the vertices of $\{a_{i,j}\}$ and $\{b_{i-1,j}\}$, we start with 10 and then alternate 011 and 010. The three successive faces at corresponding positions in two adjacent wedges contribute to the degree of a single vertex, except that the edge vertices are incident to only one face in each wedge and the next vertex to only two. Since the sum of the two sequences is $21(021)^{r-1}$, each vertex receives a contribution of 3 to raise its degree to 9, except that the vertices on the inner and outer ends of the wedges receive 2 and 1, respectively, as desired. We have added vertices to $7 + 9r$ faces, so $n = 18r + 10$. Again this construction matches the bound.

For $k = 10$, suppose $n = 21r + 11$; the construction is simpler than that for $k = 9$. Start with G_B for $m = 9r + 3$, as for $k = 9$. Add vertices to all four inner and all four outer faces, so the degrees of vertices on the wedges are now $8, 6, \ldots, 6, 8$. Within each wedge, use the sequence $11(011)^{r-1}$. The sum of the contributions from two adjacent wedges is $22(022)^{r-1}$, and each vertex of B receives the needed contribution. We have added vertices to $8 + 12r$ faces for $n = 21r + 11$, and the construction matches the bound. □

For $k = 11$, the problem is a bit more difficult, because the upper bound on the sum of the $m(n, 11)$ largest degrees is a bit larger than the maximum sum of the degree exceeding 11. We begin with the construction.

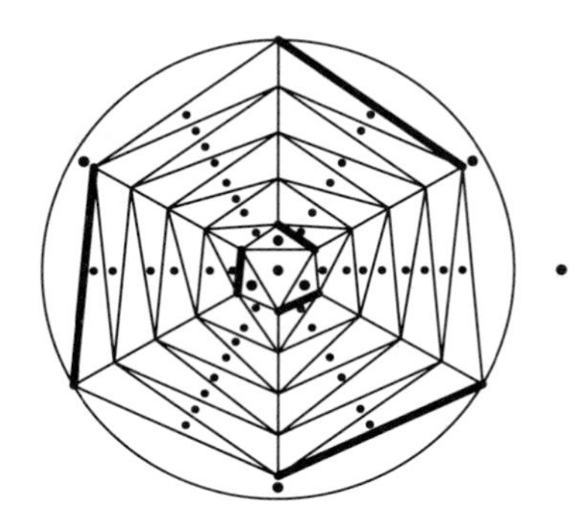

FIGURE 4.

LEMMA 3.1. *For $n = 24r + 14$ and $r \geq 1$, $f(n, 11) \geq D(n, m(n, 11)) - 6$.*

PROOF. Begin with our previous base graph with six spokes on $m = 9r + 3$ vertices, as shown in Figure 4. Add vertices to all four inner and all four outer faces, so the degrees of vertices on the wedges are now $8, 6, \ldots, 6, 8$. In the wedge formed by the vertices of $\{a_{i,j}\}$ and $\{b_{i+1,j}\}$, we use the sequence $11(111)^{r-1}$. In the wedge formed by the vertices of $\{a_{i,j}\}$ and $\{b_{i-1,j}\}$, we use the sequence $11(011)^{r-1}$. The sum of the contributions from two adjacent wedges is $22(122)^{r-1}$, and each vertex of B receives the needed extra 5 neighbors except for the innermost 6 and outermost 6 vertices which have degree only 10. To bring them to degree 11 add a vertex to each of the darkened edges $a_{i,1}b_{i-1,1}$ and $a_{i,\lceil (r+1)/2 \rceil}b_{i-1,\lfloor (r+1)/2 \rfloor}$ of G_B. We have added $8 + 3(5r - 1) + 6$ vertices for a total of $n = 24r + 14$. □

This construction fails to meet the bounds of the previous section, because it adds six vertices to edges before all faces have received vertices. Nevertheless, it is optimal, and we can improve the upper bound by six to match it. The proof of this is surprisingly long.

When we are maximizing the sum of degrees above threshold instead of the m largest vertex degrees, the proof in Lemma 2.1 that G_B is a triangulation is no longer valid. Fortunately, the first two paragraphs of the proof remain valid, and we may assume that G and G_B are connected. Instead of knowing that G_B is a triangulation, the following inequality will suffice.

LEMMA 3.2. *Let G be a connected simple plane graph with m vertices, e edges, and f faces. Let R be a specified set of vertices in G, whose intersections with the face boundaries of G have sizes $l_1 \geq l_2 \geq \ldots \geq l_f$. If $1 \leq r \leq f$, then $2e + \sum_{i=1}^{r} l_i \leq 2(3m - 6) + 3r$.*

PROOF. If G has a face of length exceeding 3, then we can increase the left side of the inequality by adding a triangular chord to such a face. Hence the left side is maximized when G is a triangulation, in which case it equals the right side. $\square$

LEMMA 3.3. *Let G be a simple n-vertex planar graph, $B = \{v \in G : d(v) \geq 11\}$, $m = |B|$, and $T = \sum_{v \in B} d(v)$. Then*

$$T \leq \begin{cases} 2n - 16 + 6m & 3 \leq m < \frac{1}{3}(n-2) \\ 3n - 18 + 3m & m \geq \frac{1}{3}(n-2) \end{cases}$$

PROOF. As already remarked, we may assume G_B is connected. Let e be the number of edges in G_B, and let r be the number of faces of G_B that contain vertices of G. The value of T is $2e$ plus the contribution from the $n-m$ vertices in $S = V(G) - B$. By Lemma 2.2, we do best by placing vertices in the r faces of G_B containing the most vertices of B (since we are not assuming G_B is 2-connected, this need not be the same as the s longest faces). Let the ith largest number of vertices on a face of G_B be l_i, and suppose the corresponding face contains s_i vertices of G. Using Lemma 2.2 to bound the contribution of edges between S and B, and then invoking Lemma 3.2, we have

$$T \leq 2e + \sum_{i=1}^{r}(l_i + 2(s_i - 1)) \leq 2(3m-6) + 3r + 2(n-m) - 2r = 2n + 4m - 12 + r$$

Since a planar graph has at most $2m-4$ faces, $r \leq 2m-4$. Also $r \leq n-m$, but we argue that at least 6 vertices of $G-B$ must be added to edges of G_B and thus $r \leq n-m-6$. To see this, let $d'(v)$ denote the degree of v in G_B, and let $U = \{v \in G_B : d'(v) < 6\}$. If we add vertices to all the faces of G_B, vertices in U still have degree at most 10 and must have a vertex added to an incident edge. If $|U| \geq 12$, then at least 6 vertices must be added to edges, since each contributes 1 to at most two vertices in U. If $U \leq 12$, then the number of added edges must be at least half of $\sum_{v \in U}(11 - 2d'(v)) = 11|U| - 2\sum_{v \in B} d'(v) + 2\sum_{v \in B-U} d(v) \geq 11|U| - 2(6m-12) + 2(6(m-|U|)) = 24 - |U| \geq 12$

We obtain the bounds claimed by plugging in these estimates for r in two cases, depending on which of $\{2m-4, n-m-6\}$ is smaller. $\square$

THEOREM 3.5. *Let G be an n-vertex planar graph, $B = \{v \in G : d(v) \geq 11\}$, $m = |B|$, and $T = \sum_{v \in B} d(v)$. Then*

$$m \leq \begin{cases} \frac{2n-16}{5} & 16 \leq n < 38 \\ \frac{3n-18}{8} & n \geq 38 \end{cases}$$

PROOF. First note that both bounds are bigger than 3, so there is nothing to prove unless $m \geq 4$. Using the bound in the last lemma and the trivial fact $11m \leq T$, we have

$$11m \leq \begin{cases} 2n - 16 + 6m & 3 \leq m < \frac{1}{3}(n-2) \\ 3n - 18 + 3m & m \geq \frac{1}{3}(n-2) \end{cases}$$

Hence whenever $3 \leq m < \frac{1}{3}(n-2)$ we have the bound $m \leq \frac{2n-16}{5}$, and whenever $m \geq \frac{1}{3}(n-2)$ we have the bound $m \leq \frac{3n-18}{8}$.

If $n < 38$ and $m \geq \frac{1}{3}(n-2)$, then the second bound says $m \leq \frac{3n-18}{8}$. Together these imply $\frac{1}{3}(n-2) \leq \frac{3n-18}{8}$, which is equivalent to $n \geq 38$ and contradicts the hypothesis. Hence we must have the first bound when $n < 38$.

If $n \geq 38$ and $m \geq \frac{1}{3}(n-2)$, then the second bound $m \leq \frac{3n-18}{8}$ applies as claimed, so suppose $m < \frac{1}{3}(n-2)$. As noted above, $n \geq 38$ is equivalent to $\frac{1}{3}(n-2) \leq \frac{3n-18}{8}$, so in this case we obtain $m \leq \frac{3n-18}{8}$ (again). $\square$

COROLLARY 3.2.

$$f(n,11) \leq \begin{cases} 2n - 16 + 6\lfloor \frac{2n-16}{5} \rfloor & \text{if } 16 \leq n < 38 \\ 3n - 12 + 3\lfloor \frac{3n-12}{8} \rfloor & \text{if } n \geq 38 \end{cases}$$

with equality for $n = 24r + 14 \geq 38$.

PROOF. Let G be an n-vertex planar graph, let $B = \{v \in V(G) : d(v) \geq 11\}$, $m = |B|$, and $T = \sum_{v \in B} d(v)$. Our lemma gives a bound on T which is monotone increasing in m, so we can obtain a bound on T by using the bound on m obtained in the last theorem. Note that $n < 38$ is equivalent to $\frac{2n-16}{5} \leq \frac{1}{3}(n-2)$. Hence in the first case the last theorem implies $m \leq \frac{2n-16}{5} \leq \frac{1}{3}(n+4)$, and then $T \leq 2n - 16 + 6m \leq 2n - 16 + 6\lfloor \frac{2n-16}{5} \rfloor$ from the previous lemma.

This leaves the case $n \geq 38$, where the last theorem gives $m \leq \frac{3n-18}{8}$. If $\frac{3n-18}{8} \geq \frac{1}{3}(n-2)$, we obtain the bound claimed for T. Because our bound for T is monotonic in m it also holds if $m < \frac{1}{3}(n-2)$. The bound is achieved by the construction in Lemma 3.1. $\square$

REFERENCES

1. Jerry Griggs presented his question at the DIMACS Workshop on Planar Graphs, November 1991. The other references to Erdős and Vince, Chung, Griggs,and Tovey were by private communication.
2. B. Grünbaum and T. S. Motzkin, *The number of hexagons and the simplicity of geodesics on certain polyhedra*, Canad. J. Math **15** (1963), 744–751.

DEPARTMENT OF MATHEMATICS, UNIVERSITY OF ILLINOIS, URBANA, ILLINOIS 61801
E-mail address: west@uiucmath.math.uiuc.edu

DEPARTMENT OF MATHEMATICS, UNIVERSITY OF ILLINOIS, URBANA, ILLINOIS 61801

List of Participants

Mike Albertson	albertson@smith.smith.edu
Noga Alon	noga@math.tau.ac.il
Dan Archdeacon	archdeac@uvm-gen.uvm.edu
Jorgen Bang-Jensen	jbj@imada.ou.dk
Lowell Beineke	beineke@cvax.ipfw.indiana.edu
Dan Bienstock	dano@cunixb.cc.columbia.edu
Endre Boros	boros@trump.rutgers.edu
Rod Canfield	erc@pollux.cs.uga.edu
Guantao Chen	gchen@plains.nodak.edu
Kiran Chilakamarri	kiran@cesvxa.ces.edu
Peter Christopher	peterrc@wpi.wpi.edu
Marek Chrobak	marek@ucrmath.ucr.edu
Fan R. K. Chung	frkc@bellcore.com
Vasek Chvatál	chvatal@cs.rutgers.edu
Nate Dean	nate@bellcore.com
Guoli Ding	ding@marais.math.lsu.edu
Dwight Duffus	dwight@mathcs.emory.edu
Genghua Fan	atfan@asuacad.bitnet
Martin Farach	farach@dimacs.rutgers.edu
Mike Fellows	mfellows@csr.uvic.ca
Zoltan Füredi	zoltan@symcom.math.uiuc.edu
Ron Graham	rlg@research.att.com
Jerry Griggs	griggs@math.scarolina.edu
Ruth Haas	rhaas@smith.smith.edu
Glenn Hurlbert	hurlbert@calvin.la.asu.edu
David Johnson	dsj@research.att.com
Jeff Kahn	jkahn@math.rutgers.edu
Howard Karloff	howard@cc.gatech.edu
Mark Kayll	kayll@math.rutgers.edu

Suh-Ryung Kim	kims@sjuvm.bitnet
Daniel J. Kleitman	djk@math.mit.edu
Tom Leighton	ftl@math.mit.edu
Charles Little	C.little@massey.ac.nz
Anna Lubiw	alubiw@maytag.waterloo.edu
Joe Malkevitch	joeyc@cunyvm.cuny.edu
Gary Miller	glmiller+@abacus.theory.cs.cmu.edu
Bohan Mohar	bojan.mohar@uni-lj.ac.mail.si
Seffi Naor	naor@cs.technion.ac.il
Jarik Nešetřil	nesetril@cspug11.bitnet
Laszlo Pyber	h1130pyb@ella.hu
Vijaya Ramachandran	vlr@cs.utexas.edu
Bruce Reed	breed@muff.cs.mcgill.ca
Fred Roberts	froberts@dimacs.rutgers.edu
Neil Robertson	robertso@function.mps.ohio-state
Bruce Rothschild	blr@math.ucla.edu
Mike Saks	saks@math.rutgers.edu
Ed Scheinerman	ers@cs.jhu.edu
Dick Schelp	schelpr@hermes.msci.memst.edu
Walter Schnyder	wschnyder@eagle.wesleyan.edu
Lex Schrijver	lex@cwi.nl
Paul Seymour	pds@bellcore.com
Greg Shannon	shannon@luap.cs.indiana.edu
Alistair Sinclair	sinclair@icsi.berkeley.edu
Diane Souvaine	dls@cs.rutgers.edu
Gabor Tardos	h679tar@ella.hu
Bob Tarjan	ret@princeton.edu
Prasad Tetali	prasad@research.att.com
Robin Thomas	thomas@math.gatech.edu
Ann Trenk	atrenk@lucy.wellesley.edu
Tom Trotter	wtt@bellcore.com
Dirk Vertigan	vertigan@dimacs.rutgers.edu
Chi Wang	c0wang01@ulkyvx.bitner
Doug West	west@uiucmath.math.uiuc.edu
Gill Williamson	gwilliamson@odin.ucsd.edu
Peter Winkler	pw@bellcore.com
C. Q. Zhang	cqzhang@wvnvm.wvnet.edu